International Political Economy Series

General Editor: **Timothy M. Shaw**, Professor of Commonwealth Governance and Development, and Director of the Institute of Commonwealth Studies, School of Advanced Study, University of London

Titles include:

Susan M. McMillan
FOREIGN DIRECT INVESTMENT IN THREE REGIONS OF THE SOUTH AT THE END
OF THE TWENTIETH CENTURY

James H. Mittleman and Mustapha Pasha (*editors*)
OUT FROM UNDERDEVELOPMENT
Prospects for the Third World (Second Edition)

Lars Rudebeck, Olle Törnquist and Virgilio Rojas (*editors*)
DEMOCRATIZATION IN THE THIRD WORLD
Concrete Cases in Comparative and Theoretical Perspective

Howard Stein (*editor*)
ASIAN INDUSTRIALIZATION AND AFRICA
Studies in Policy Alternatives to Structural Adjustment

International Political Economy Series
Series Standing Order ISBN 978-0-333-71708-0 hardcover
Series Standing Order ISBN 978-0-333-71110-1 paperback
(*outside North America only*)

You can receive future titles in this series as they are published by placing a standing order. Please contact your bookseller or, in case of difficulty, write to us at the address below with your name and address, the title of the series and one of the ISBNs quoted above.

Customer Services Department, Macmillan Distribution Ltd, Houndmills, Basingstoke, Hampshire RG21 6XS, England

Innovation and Social Learning

Institutional Adaptation in an Era of Technological Change

Edited by

Meric S. Gertler
Professor of Geography and Goldring Chair in Canadian Studies
University of Toronto
Canada

and

David A. Wolfe
Professor of Political Science
University of Toronto
Canada

Published by
PALGRAVE MACMILLAN
Houndmills, Basingstoke, Hampshire RG21 6XS and
175 Fifth Avenue, New York, N. Y. 10010
Companies and representatives throughout the world

PALGRAVE MACMILLAN is the global academic imprint of the Palgrave
Macmillan division of St. Martin's Press, LLC and of Palgrave Macmillan Ltd.
Macmillan® is a registered trademark in the United States, United Kingdom
and other countries. Palgrave is a registered trademark in the European
Union and other countries.

ISBN-13: 978–0–333–75284–5

This book is printed on paper suitable for recycling and
made from fully managed and sustained forest sources.

A catalogue record for this book is available from the British Library.

Library of Congress Catalog Card Number: 2001058212

For Joanna and Lisa

Contents

List of Tables

Acknowledgments

Many of the authors gathered in this volume first met to discuss these issues at a seminar held in Toronto in May, 1994 on 'Institutions of the New Economy'. This seminal event was organized by Professor Liora Salter of Osgoode Hall Law School under the auspices of the Program on Law and the Determinants of Social Ordering sponsored by the Canadian Institute for Advanced Research. We are deeply indebted to Liora for her leadership in providing the initial stimulus for the collection of papers presented here; as well as her continuing friendship and intellectual support throughout the development of the project.

We would also like to acknowledge the excellent assistance we received from a small, but dedicated, group of research associates in the preparation and editing of this collection. We are especially indebted to Lisa Mills for keeping the project on the rails at several key moments along the way; and more recently, to Matthew Lucas and Norma Rantisi for their crucial editorial and organizational input. We would also like to thank the University of Toronto's Centre for International Studies for its institutional support of this project, as well as the staff and colleagues at the Munk Centre for International Studies who have provided us with such a productive and welcoming milieu in which to work.

Meric Gertler would like to acknowledge the Social Sciences and Humanities Research Council of Canada for its generous support of his research program and the work presented in this volume.

David Wolfe would also like to thank the Canadian Institute for Advanced Research for a research associateship which afforded him the opportunity to pursue the research that underlies his contributions to the volume, and in particular, its Founding President, and current Fellow of the Institute, Dr Fraser Mustard, who over the past decade has successfully integrated the concepts of learning and innovation in all of his considerable undertakings.

List of Abbreviations

ASEAN	Association of Southeast Asian Nations
BIS	Bank for International Settlements
BIT	Bilateral Investment Treaty
BW	Baden-Württemberg
C-R	Conventional or Relational Transactions
CAP	Common Agriculture Policy
CEE	Central and Eastern European
CIS	Community Innovation Survey
CSE	Consumer Subsidy Equivalent
CUFTA	Canada–US Free Trade Agreement
ECMA	European Computer Manufacturers Association
ERDF	European Regional Development Fund
ESPRIT	European Strategic Program for Research and Development on Information Technology
ETSI	European Telecommunications Informatics Services
EU	European Union
EUREKA	European Research Coordination Agency
FAO	Food and Agriculture Organization
FATF	Financial Action Task Force on Money Laundering
FCC	Federal Communications Commission
FDI	Foreign Direct Investment
FIBV	International Federation of Stock Exchanges (English translation)
FP	Framework Program
FRS or LFR	Less-favored Region
G-7	Group of Seven
G-10	Group of Ten
G-30	Group of Thirty
GATS	General Agreement on Trade in Services
GATT	General Agreement on Tariffs and Trade
GDP	Gross Domestic Product
ICT	Information and Communication Technologies
IMF	International Monetary Fund
IN	Indiana
IOSCO	International Organization of Securities Commissions
ITU	International Telecommunications Union
KY	Kentucky
LDCs	Less Developed Countries
LFR	Less-favored Region

M&As	Mergers and Acquisitions
MAI	Multilateral Agreement on Investment
MFN	Most-favored nation
MI	Michigan
MITI	Ministry of International Trade and Industry
MITT	Ministry of Industry, Trade and Technology
NAFTA	North American Free Trade Agreement
NDP	New Democratic Party
NGOs	Non-governmental Organizations
OECD	Organization for Economic Cooperation and Development
PBTL	Product-based Technological Learning
PSE	Producer Subsidy Equivalent
R&D	Research and Development
REGIS	Regional Innovation Systems: Designing for the Future
RIDBS	Report on International Developments in Banking Supervision
RIS	Regional Innovation Strategy
RP	Reference Paper
RTD	Research and Technology Development
RTP	Regional Technology Plan
SDO	Standard Developing Organization
SI	Systems of Innovation
SME	Small and Medium-sized Enterprise
SPF	Sector Partnership Fund
STRIDE	Science and Technology for Regional Innovation in Europe
TGV	Train à Grande Vitesse
TN	Tennessee
TQM	Total Quality Management
TRIMs	Trade Related Investment Measures
UK	United Kingdom
UN	United Nations
UNCTAD	United Nations Conference on Trade and Development
UNDP	United Nations Development Program
US	United States
WDA	Welsh Development Agency
WTO	World Trade Organization

Notes on the Contributors

Philip Cooke is Professor of Regional Development and Director of the Centre for Advanced Studies at the University of Wales, Cardiff. He is the author of numerous articles on regional innovation systems, inter-firm networks and regional–global interactions in economic affairs. His most recent books are: *The Governance of Innovation in Europe* (1999, with P. Boekholt and F. Toedtling), *The Associational Economy* (1998, with K. Morgan), and *Regional Innovation Systems* (1998, co-edited with H. Braczyk and M. Heidenreich). He has served as advisor to the Science and Technology Directorate of the EU, the UK Department of Trade and Industry, the governments of Ireland, Norway, and numerous regional administrations on business clustering and regional development.

Richard Florida is H. John Heinz III Professor of Regional Economic Development at Carnegie Mellon University. He is the author of *The Rise of the Creative Class: and How It's Transforming Work, Leisure, Community and Everyday Life* (2002).

Meric S. Gertler is Professor of Geography and Goldring Chair in Canadian Studies at the University of Toronto. He is also co-Director of the Program on Globalization and Regional Innovation Systems at the University of Toronto's Center for International Studies. Together with David Wolfe, he co-directs the Innovation Systems Research Network, a national network of scholars funded by the research councils of the Government of Canada. He has served as consultant to the OECD, the European Commission and various local, provincial and federal agencies in Canada. Among his recent publications are: *The New Industrial Geography: Regions, Regulation and Institutions* (with T. Barnes (1999)) and *The Oxford Handbook of Economic Geography* (with G.L. Clark and M.P. Feldman (2000)).

Dylan Henderson is a Research Associate in the Department of City and Regional Planning, Cardiff University. He has recently completed his PhD on Regional Innovation Networks in Wales, focusing in particular on the design and implementation of the Welsh Technology Plan.

Kevin Morgan is Professor of European Regional Development in the Department of City and Regional Planning at Cardiff University. His research interests include regional innovation strategies in Europe and the relationship between devolved governance structures, democracy and development. He is the co-author (with Philip Cooke) of *The Associational Economy: Firms,*

Regions and Innovation (1998) and co-editor (with C. Nauwelaers) of *Regional Innovation Strategies: the Challenge for Less Favoured Regions* (1999).

Lynn K. Mytelka is a Professor in the Department of Political Science at Carleton University and Director of the Institute for New Technologies (UNU/Intech) at the United Nations University. From 1996 to 1999, she was Director of the Division on Investment, Technology and Enterprise Development at the United Nations Conference on Trade and Development (UNCTAD). She is the editor of *Competition, Innovation and Competitiveness in Developing Countries* (1999), and, with Dieter Ernst and Tom Ganiatsos, editor of *Technological Capabilities and Export Success in Asia* (1998).

Tony Porter is Associate Professor at the Department of Political Science, McMaster University. He is currently involved in several research projects including: the role of private international institutions in international regimes; the relationship of private institutions, states and technologies in the governance of international industries; and the governance of global finance. He is co-editor (with A. Claire Cutler and Virginia Haufler) of *Private Authority in International Affairs* (1999), and author of *States, Markets and Regimes in Global Finance* (1993).

Liora Salter is a Professor of Law at Osgoode Hall Law School, and of Environment Studies at York University. She has completed a multi-year study of standard setting, and has a long-term interest in issues of regulation, public policy, science policy, and participation. Her books on these topics include *Public Inquiries in Canada* (1981), *Managing Technology* (1990), *Mandated Science* (1988), and a special issue of the *International Journal of Political Economy*. She also writes on communication issues. She is a fellow of the Royal Society of Canada.

Michael Storper is Professor of Urban Planning at the University of California at Los Angeles and Professor of Human and Social Sciences at the Université de Marne-la-Vallée. Among his most recent publications are *The Regional World: Territorial Development in a Global Economy* (1997), and, with S. Thomadakis and L. Tsipouri, *Latecomers in the Global Economy* (1998).

David A. Wolfe is Professor of Political Science at the University of Toronto and Co-Director of the Program on Globalization and Regional Innovation Systems (PROGRIS) at the Munk Centre for International Studies. PROGRIS is the node for one of five subnetworks of the Innovation Systems Research Network (ISRN), funded by the Social Sciences and Humanities Research Council of Canada, and serves as the national secretariat for the network. Recent publications include *Innovation, Institutions and Territory: Regional Innovation Systems in Canada* (2000, co-edited with J. Adam Holbrook) and

Knowledge, Clusters and Regional Innovation: Economic Development in Canada (2002, co-edited with J. Adam Holbrook).

Robert Wolfe is Professor of Policy Studies, Queen's University, Kingston, Ontario. He is the author of *Farm Wars: the Political Economy of Agriculture and the International Trade Regime* (1998).

1
Innovation and Social Learning: an Introduction

David A. Wolfe and Meric S. Gertler

Introduction

We live in an era of rapid economic and technological change marked by a high degree of uncertainty. Over the past ten years, we have witnessed the shift from a steep recession at the beginning of the decade through two international financial crises to the recent phase of hyper growth, marked by higher levels of employment, significant gains in productivity and a rise in consumer spending, especially in North America. So strong has been the economic performance of the past few years that some commentators have hailed it as the return to a 'golden age', similar to that which marked the three decades following the Second World War – despite the economic slowdown beginning in late 2000. This shift is attributed to two underlying developments: the first is the trend towards globalization which is increasing the linkages between Europe, North America and East Asia in terms of investment, trade, research and development; and the second is the emergence of an integrated set of information and communication technologies linking diverse communications and entertainment media together in digital form. Together, these trends are reshaping the economies of the industrial and industrializing economies and changing much of the accepted wisdom about how they operate.

Despite this nascent optimism about the future, many still fear a return to the conditions of slower growth and higher unemployment of the preceding decades. Even in those countries currently enjoying their new-found prosperity, considerable uncertainty remains about how long it will last and concerns are being raised about the inequitable distribution of gains and losses from the transition to this 'new economy'. The US government has drawn attention to the emergence of a growing 'digital divide' within its society between those who enjoy access to, and the benefits of, the new technologies, and those who face exclusion from their share of the benefits (US Dept. of Commerce, 2000). In Europe, similar concerns have been expressed about the dangers of a Europe moving at two speeds – in both a social and a geographic

1

sense. Over the past decade, considerable energy has been devoted to ensuring that the less favoured regions of the continent share equally in the prospects for growth that a more integrated market is expected to bring (Morgan and Nauwelaers, 1999).

The ability of individual economies and societies to respond to the challenge of the current transition is determined, in large measure, by the capacity of existing institutions to adapt to the changes underway. While the nature and extent of the adaptation required may seem novel, the process itself is not. Periods of rapid economic and technological change are characterized by a condition of extreme uncertainty. They place a premium on the ability to acquire, absorb and diffuse relevant knowledge and information throughout the various institutions that affect the process of economic development and change. Two key concepts appear to be central to the concerns at the heart of this question: the role of institutions and the process of learning. The emphasis on institutions emerges out of the ongoing insights generated by the field of evolutionary economics which emphasizes that innovation is increasingly a social enterprise that occurs within a variety of institutional settings: primarily the firm, but also the other institutions that comprise the innovation system. In this sense it is imperative to develop an understanding of the dynamics within institutional settings that impact on, foster or constrain the innovation process. Second, the centrality of learning for the innovation process stems from the recognition that the knowledge frontier is moving so rapidly in the current economy that simple access to, or control over, knowledge assets affords merely a fleeting competitive advantage. It is the capacity to learn which is critical to the innovation process and essential for developing and maintaining a sustainable competitive advantage.

The focus on institutions arises from the simple observation that virtually all economic activity occurs within an institutional context. In the economic analyses of Weber, Veblen, Schumpeter and Karl Polanyi, economic processes are embedded and enmeshed in a variety of institutions, including habits and customs, as well as government, religion, culture and the legal framework of a society. For Polanyi, in particular, the instituting of economic processes endows them with a unity and stability by creating a structure that has a definite function in different societies; however, the specific way that economic processes are embedded in both the economic and noneconomic institutions of a society varies (Polanyi, 1957). 'Markets do not exist or operate apart from the rules and institutions that establish them and that structure how buying, selling and the very organization of production take place' (Zysman, 1994, p. 244). This concept of institutions also bears a close affinity to that found in the neo-institutionalist stream of the international relations literature. From that perspective, institutions reflect persistent and connected sets of rules, formal and informal, that prescribe behavioral roles, constrain activity and shape expectations (Ruggie, 1982; Keohane, 1990).

The critical issue is: how well- or ill-suited are the institutions of a region, nation or international regime to the task of coping with the magnitude of change currently underway in the global economy? In an economy where information is becoming an increasingly fundamental commodity and the very basis of production is becoming more knowledge-intensive, the role of institutions in retaining and transmitting knowledge to their members becomes ever more crucial. The emerging digital economy places greater emphasis on the importance of knowledge in all areas of activity, making it vital that the knowledge being socially stored and transmitted is appropriate to the emerging technological style or paradigm.

Closely related to this is the question of how social learning actually occurs in an institutional context. Economists concerned with questions of technological change have identified a range of mechanisms through which institutional learning occurs: learning-by-doing, learning-by-using and learning-by-interacting. The majority of this literature focuses on the narrower and, to some extent, more conventional issues of producing and disseminating technological knowledge. The emphasis has largely been on the process of learning by searching that is intrinsic to innovation, the increasing importance of cooperative relations between users and producers of technology or networks of producers in disseminating new knowledge, or the role of specialized educational institutions in providing necessary supports to the innovation process (Lundvall, 1988; Johnson, 1992).

However, the broader issues raised by the current period of social change require a more inclusive notion of social learning – one which focuses on the capacity of institutions to sustain growth and facilitate the adjustment process from declining sectors and occupations to expanding ones by adapting and changing in response to new competitive conditions. This particular concept of learning is critical for the kinds of organizational changes associated with the emerging, knowledge-based economy. Increasingly, the organizational issue is how to pool and structure knowledge and intelligence in social ways, rather than access them on an individual basis. The capacity for social learning and increased networking may be seen as essential for tapping into the shared intelligence of both the individual firm or organization, as well as a collectivity of firms within a given geographic space. This form of shared or networked learning assumes that neither the public sector nor individual private enterprises are the source of all wisdom; rather, the process of innovation and institutional adaptation is essentially an interactive one in which the means for establishing supportive social relations and of communicating insights and knowledge in all its various forms are crucial to the outcomes.

This insight suggests a higher order of learning by institutions – one based on a capacity for reflexivity and the ability to apply institutional memory and intelligence to monitor the success of institutions in adapting to ongoing changes in the environment. This higher order is *learning-by-learning* where the (institutional) self-monitoring of the learning process itself becomes an integral feature of the institutional structure. Whether the organized intelligence and institutional adaptability of a region, nation or international regime constitutes a progressive force for change in the process of restructuring or an 'institutional drag' depends on this self-monitoring, or the ability to shed inefficient norms and practices and replace them with ones that facilitate the process of economic change and social adaptation (Sabel, 1994; Cooke, 1997).

This higher level of institutional learning is potentially relevant for regional, national, and supranational levels of governance. The forces described above are not only affecting individual economies and societies, but also altering the relations between the national, supranational and regional levels of governance. The spread of globalization and the growing interdependence of individual economies have led to a debate about the capacity of national institutions to respond to the changes underway in the global economy. Some argue that the growing disjuncture between the formal authority of the nation state and the emerging global system of production and distribution is shifting attention away from the national level to the supranational and subnational levels. Others are not as quick to accept the demise of national institutions in the governance of innovation-based learning. Nevertheless, developments associated with the trend towards globalization reinforce the growing salience of supranational institutions: the internationalization of production and of financial markets; the integrative capacity of information technologies that overcome many of the previous economic barriers in transportation and communications; the increased power of international regimes and organizations in the management of economic affairs; and the increasing scope of power and authority delegated upward to supranational bodies. Collectively, these trends are contributing to an increased focus on the institutional capacities of emerging international organizations and supranational bodies.

Conversely, the impact of new technologies also focuses attention on the role of regions. A number of factors contribute to the increasing prominence of regions in the process of institutional change and social learning. Geographers have long noted that complex systems of technology, production processes, industrial organization and their supporting infrastructures of social and political institutions, exhibit distinctive spatial characteristics. Production relations tend to aggregate over time among networks of firms following the pattern of input–output relations, or traded interdependencies, that provide the basis for knowledge exchange in the local economy. These technological spillovers are tied to knowledge and practices that are often tacit (and hence, context-specific), rather than explicit (Cooke and Morgan, 1998). Together, the forms of collaboration and interaction associated with both traded and untraded interdependencies highlight the importance of the regional dimension, especially in the current era of rapid technological change and increasing globalization (Storper, 1997).

The role of institutions

Before proceeding further, it is important to have a clear sense of the nature of institutions themselves, and the role they play in regulating economic life. There are a number of different approaches to the concept of institutions in sociology and economics, encompassing a rich and well-established intellectual legacy. At their broadest level, institutions incorporate social roles based on established norms and expected patterns of behavior, thus precluding the necessity for individuals to relearn their social roles every day. They operate as an important mechanism for transmitting information about accepted norms and expected patterns of behavior to the members of society. The origins of this idea go back to the early part of the twentieth century, to the writings of Max Weber, who introduced the fundamental concept of *social action*. Social action is that action whose subjective meaning takes account of the behavior of others and is thereby oriented in its course. Many types of social action are guided by the belief in the existence of a legitimate order. Legitimate orders in turn are guaranteed in two ways: 1) through purely subjective means based on affect, belief in an ethical or esthetic set of values, or religious belief; and 2) by the expectation of specific external effects, or interest situations. Weber distinguishes between two such sets of effects: 1) convention, where deviation from it within a social group results in a reaction of disapproval; and 2) law, where there is a high probability that physical or psychological coercion will be applied in order to ensure compliance or avenge violation. Finally, he defines a social relationship whose regulations are enforced by specific individuals, usually a chief and an administrative staff, as an organization (Weber, 1978, pp. 4–34).

Weber's typologies of social action and legitimate orders have been adopted and used in many different contexts, but one of the most enduring is that established by Hans Gerth and C. Wright Mills (1954). While clearly acknowledging their debt to Weber, Gerth and Mills build into their conceptual framework the insights of social psychology, which adds a dimension missing from Weber. Where social action constitutes the basic unit of analysis for Weber, Gerth and Mills adopt the concept of *role* as theirs. This concept 'refers to: 1) units of conduct which by their recurrence stand out as regularities and 2) which are oriented to the conduct of other actors. These recurrent interactions form patterns of mutually oriented conduct' (Gerth and Mills, 1954, p. 10). The roles played by people are delimited by the kind of social institutions into which they are born and mature. A person may play many different roles and each of these may be a segment of the different institutions and situations through which they move. An institution is seen as an organization of roles which carry different degrees of authority, so that one of the roles is understood and accepted as guaranteeing the relative permanence of the total conduct pattern (Gerth and Mills, 1954, p. 13).

Following in this tradition, Berger and Luckmann's classic treatise assigns institutions and the process of institutionalization a central role. According to them, 'institutionalization occurs whenever there is a reciprocal typification of habitualized actions by types of actors' (Berger and Luckmann, 1967, p. 54). The reciprocal typifications of actions that constitute institutions always take place in the context of a shared history. The nature of the institution cannot be understood without a knowledge of the historical process in which it was produced. Institutions also channel and define human conduct by subsuming it under social control.

From the perspective of evolutionary economics, institutions occupy a similar status, but they play specific roles in the functioning of an economy. They reduce uncertainty in everyday life by forming patterns of interaction and shaping the way individuals view and understand society. Institutions are central to the process of learning discussed above. Learning processes are inherently social and interactive, not just individual, and new knowledge is created through processes that are institutionally embedded. Institutions also provide basic functions for the operation of economies.

> They provide information, reduce uncertainty, manage conflicts and cooperation, and create incentives and trust. These functions not only give stability and structure to the economy, they are also crucially important for innovation. All innovative activities are riddled with uncertainty and in the modern economy there are many institutions to assist in coping with the technical and financial uncertainties of innovation (Johnson and Nielsen, 1998, pp. xiii–xv).

This institutionalist perspective is grounded in an older tradition in economics closely associated with the work of Thorstein Veblen and John R. Commons. While there are important differences within this tradition, the work of Veblen has the greatest relevance to the present undertaking. Veblen was particularly concerned with investigating the effects of technological change on institutional structures and with the ways in which established social conventions resist such change. Veblen's focus on institutions shares in common key concerns with the sociological tradition, in its emphasis on the importance of viewing human action within the context of its institutional surroundings. For Veblen, technology is the driving force of economic change. Its pace and direction are affected by the institutional framework within which it occurs. It also has institutional consequences in the way in which it alters the material circumstances of individuals and the methods, patterns and habits of them as well (Rutherford, 1994, pp. 38–9). Veblen also discusses the ways in which material and technological conditions which shape patterns of life subsequently become subject to convention or part of our commonly held values and beliefs – what Veblen called 'settled habits of thought' (Veblen, 1919, p. 239).

An important parallel to these themes in the work of Veblen is found in the thinking of Harold Innis. Innis, who studied at the University of Chicago but after Veblen's time, was nonetheless strongly influenced by Veblen's institutionalism (Barnes, 1999). Running through his work is a pre-occupation with questions of the techniques of production of particular commodities or forms of communication and the broader institutional structures and social relationships, which grow up around and sustain those forms of production. In this regard, Innis was concerned with questions of stability and instability in social and economic relations and with the factors that disturb that stability. In his desire to understand the factors that divert a pattern of economic relations from a path of balanced growth, he devoted considerable attention to disruptive factors, such as the introduction of new technologies; to unstable factors, such as the persistence of massive overhead costs and 'unused capacity'; and to rigidities, such as the exercise of monopoly power.

In his perceptive essay on 'The Penetrative Powers of the Price System', the implications of some of these themes are suggested in an intriguing, but underdeveloped fashion. Drawing on the work of Sombart and Geddes, Innis noted the significance of the shift from the commercial phase of capitalism to the industrial phase and within the latter, from the paleotechnic (based on coal and iron) to the beginnings of the neotechnic (new sources of power and base metals). In a noteworthy passage, he documented the disruptive impact of new transportation technologies (the internal combustion engine) and sources of power (oil and hydroelectricity) on prevailing institutional patterns and relationships, particularly the role of the state in economic regulation. He saw the decline of the old and the rise of the new industrial technologies as a key development during the years of the Depression. Concomitantly, the geographic areas tied to the old or new technologies suffered a more or less positive fate. Most critical, however, were the implications of this transition for prevailing institutional structures and patterns of social relations:

> Neotechnic industrialism superimposed on palaeotechnic industrialism involved changes of tremendous implication to modern society and brought strains of great severity. The institutional structure built up on iron and steel and coal has been slow to change. Governmental machinery in those regions in which palaeotechnic society developed late has been extended and government intervention in regions in which it developed earlier has been intensified as a result of the rigidities of labour organization and corporate finance (Innis, 1956, pp. 263–4).

Alongside Veblen and Innis, the writings of Karl Polanyi (as noted earlier) mark an important contribution to the study of institutions. Both in his classic study *The Great Transformation* (1944) and in his more anthropological writings, Polanyi stressed that all economic relationships were grounded in institutional processes, not just the transactions of atomized individuals. For Polanyi, economies represent 'instituted processes of interaction between man [*sic*] and his environment, which results in a continuous supply of want satisfying material means' (Polanyi, 1957, p. 248). The market represented just one of the many institutional forms that economic processes could adopt. What was critical always was the specific way in which these economic processes were embedded in both economic and non-economic institutions. Religion and government could play just as important a role in establishing the structure of economic institutions in certain societies as purely private mechanisms. Nowhere did he see this as more true than in the creation of the market. Ultimately for Polanyi, the market is a socially constructed economic institution, which differs from previous forms in the extent to which it subjects other social institutions to purely economic calculations and processes: the 'self-regulating market demands nothing less than the institutional separation of society into an economic and political sphere . . . Such an institutional pattern could not function unless society was somehow subordinated to its requirements' (1957, p. 71).

The writings of Joseph Schumpeter offer a final strand of thought within the institutionalist tradition of economics. Although usually identified by his views on the dynamic role of the entrepreneur in promoting economic and technological innovation, there is a subtheme running though Schumpeter's writings that reflects a concern with the link between technological and institutional change. In contrast to Veblen, who saw existing institutions as a fetter on the prospects for economic and technological advancement, Schumpeter believed that capitalism's subversion of pre-capitalist social institutions posed the greatest threat to its own survival. Through the extension of the market and the universalization of commodity exchange, capitalism eliminated most of the institutional fetters of the old feudal order, thus destroying the protective shell of the status and political order within which it had been nurtured. Eventually, both the economic dynamics of capitalism and its socializing effects did away with the very class – the aristocracy – that had provided political leadership for the new economic order. In Schumpeter's view, its demise eliminated the strong bulwark afforded to capitalism by its feudal shell. That process, impressive in its relentless necessity, was not merely a matter of removing institutional deadwood, but of removing partners of the capitalist stratum, symbiosis with whom was an essential element of the capitalist schema (Schumpeter, 1950, p. 139).

This focus on institutions was submerged in the 1960s and 1970s under a wave of alternative approaches; but recently has experienced a renewed interest. Three approaches are helpful for the present discussion: in comparative politics, international relations, and some elements of the new institutional economics. Within the discipline of comparative politics, the work of Peter Hall highlights the need for an institutional approach to understanding the role of the state. Hall views the state as a network of institutions, operating within a complex of related institutions that form part of society and the economy. This approach locates the determining factors behind economic policy and performance in the organizational structures of the state and society. Institutions 'refer to the formal rules, compliance procedures, and standard operating practices that structure the relationship between individuals in various units of the polity and economy' (1986, p. 19). The strength of such analyses depends on the capacity of institutions to endure and adapt. To the degree that national institutions change, the key challenge is to identify the socioeconomic and political coalitions that support the change and analyze how they contribute to the process (Hall, 1997, p. 183).

This theme finds strong expression in the work of John Zysman who argues that distinctive institutional structures across nations, which are the product of historically conditioned political and industrial development, define the choices available to individual firms or actors in responding to new economic or technological trends. Historically conditioned and nationally specific institutional structures create distinctive patterns of constraints and incentives. He also links this institutional approach to the question of power by arguing that relations among institutions embed and channel the existing power relations among groups in society (1996, pp. 414–15). More recently, this emphasis on distinctive national constellations of institutional structures has been adopted with renewed enthusiasm in the emerging literature on national business systems and 'divergent capitalisms' (Doremus *et al.*, 1998; Whitley, 1999; Gertler, 2001; Hall and Soskice, 2001).

Another important source of thinking in the revival of institutional analysis has been the field of international relations (IR) where the work of Stephen Krasner, John Ruggie and Robert Keohane among others has led to a growing interest in how long-term international relations may be structured in a stable manner through the creation of international regimes without the dominance of a single hegemon. The institutionalist approach in IR theory has been contrasted with the realist approach which has long held that economic or military power is the determining factor in structuring the prevailing pattern of international relations. According to Krasner and Ruggie, international regimes are a form of social institution that constrain or influence the expectations of actors in a given area of international relations (Ruggie, 1982, p. 196). International regimes limit the ability of their members to act within the domain of that regime. The concept of institutionalism has been further expanded in the work of Robert Keohane, who delimits the institutional approach in IR theory to include those authors who see cooperation as a key feature of economic interdependence. For Keohane, the existence of shared economic interests leads to a situation where states will work together to create international institutions and rules. In a manner similar to

Berger and Luckmann, Polanyi, Hall, and Zysman, he defines institutions more specifically as 'persistent and connected sets of rules, formal and informal, that prescribe behavioural roles, constrain activity and shape expectations' (1990, p. 732). Institutionalism has also experienced a rebirth of interest in economics. Drawing inspiration from the work of Veblen, Polanyi and other institutionalists, a new generation of scholars sees the institutional approach as a necessary counterweight to the methodological individualism of the neoclassical approach. While there are many different strands that can be associated with the new institutional economics, one of the most comprehensive approaches is presented by Geoffrey Hodgson. Referring back to Veblen's conception of institutions as 'settled habits of thought', he defines institutions as 'a social organization which, through the operation of tradition, custom or legal constraint, tends to create durable and routinized patterns of behaviour' (1988, p. 10). The significance of institutions is their ability, in face of uncertainty, to create stable patterns of expectations through the incorporation of habits and routines. The use of habits and routines helps actors to deal with the complexity of everyday economic life by reducing the need for rational calculations involving large amounts of diverse information. They are also essential for incorporating the acquired skills, tacit knowledge and accumulated information of collectivities of workers organized into firms. Habits and routines organized into institutional forms within the firm thus constitute an important mechanism by which skills and technological learning are preserved and transmitted within the firm (Hodgson, 1993, p. 234).

A critical point of debate in this recent approach to institutional analysis concerns the relations between institutions and organizations. Edquist and Johnson maintain that institutions ('things that pattern behavior' such as norms, routines, rules and laws) and organizations (consciously created 'concrete things' such as firms, universities, private research labs, and technology transfer organizations) are quite distinct and that, for the sake of conceptual and empirical clarity, they should be regarded as separate (Edquist and Johnson, 1997, p. 43). They therefore argue that institutions should be defined more narrowly, akin to the approach taken by North (1990) and Williamson (1985), to mean 'sets of common habits, routines, established practices, rules, or laws that regulate the relations and interactions between individuals and groups' (Edquist and Johnson, 1997, p. 46). They proceed to demonstrate how institutions, so defined, act to manage conflicts and cooperation, provide incentives to economic actors, channel resources to innovation-generating activities and, at times, constitute obstacles to innovation.

However, recent work in the sociology of organizations disagrees with this distinction and argues that formal organizations are subsumed under the broader category of institutions. For writers such as DiMaggio and Powell, the rules and norms which constitute institutions, are reflected in organizational structures and processes. This perspective assumes that changes in organizational structures and processes reflect changes in the broader sets of institutional norms and rules in which they are embedded (1991). In his recent overview of the implications of institutions for the study of innovation, Rogers Hollingsworth explicitly endorses the view that 'institutional rules, norms and conventions unfold in tandem with organizational structures' (2000, p. 619), although these represent merely two out of what he sees as five levels of reality that are necessary for an adequate analysis of institutions. This approach is also consistent with the perspective taken by other writers in the innovation systems perspective, such as Richard Nelson, who tend to use institutions and organizations interchangeably (1988; 1994). We follow the latter approach, accepting the view that organizations are best understood as being nested within (and shaped by) the broader institutional environment in which they are situated.

Institutions and social learning in the new economy

The role of institutions and institutional change assumes a position of critical importance in periods of rapid economic and technological change. The critical concern of the essays collected in this volume is to understand how institutions inhibit or promote the process of restructuring endemic to a period of rapid economic change. As we noted at the outset, at the heart of the current transition are the dual processes of globalization and the convergence of an integrated set of computer, communications, and video technologies with the capacity to process and transmit data in digital form. The new information and communications technologies are exponentially increasing the capacity to handle information, as knowledge-based inputs simultaneously become a salient component of every aspect of production. New information technologies augment both the existing knowledge base and the need to access it in ways that demand a new capacity for learning and the absorption of knowledge. This dilemma draws attention to the fact that the quantitative change in technologies is gradually leading to a qualitative one – the emergence of 'socially distributed cognitive intelligence', or 'networked intelligence'. According to one informed assessment of this phenomenon:

> The Age of Networked Intelligence is . . . not an age of smart machines but of humans who through networks can combine their intelligence, knowledge, and creativity for breakthroughs in the creation of wealth and social development. It is not just an age of linking computers but of internetworking human ingenuity (Tapscott, 1996, p. xiv).

This point underscores the fact that economic and technological changes do not occur in isolation from underlying social and institutional transformations. The new information and communications technologies constitute the key factor, or core technology, underlying the emergence of a new techno-economic paradigm. The emergence of a techno-economic paradigm differs from less pervasive forms of innovations in terms of the characteristics associated with its key factor: a relatively low and constantly falling cost curve, plentiful source of supply, and ease of application across many sectors of the economy. However, according to Freeman and Perez, the key factor diffuses throughout a modern economy as the core of a rapidly growing system of technical, social and managerial innovations (1988, pp. 58–61). The outcome of such a transition depends on a complex process of change in forms of social organization and the resolution of political conflict. In passages reminiscent of Veblen and Innis, Perez (1983) argues that the new constellation of technologies cannot be generalized throughout an economy without a corresponding shift in a wide range of social and political institutions. Long periods of growth and decline in industrial economies result from the measure of complementarity, or lack thereof, between the prevailing organization of the production process and the dynamics of the sociopolitical institutional structure. New technological systems emerge in the sphere of production as a complex of interrelated technologies and a new form of the organization of production. The problem of adjustment emerges out of the fact that the collection of institutions that comprise the prevailing sociopolitical infrastructure no longer complements the new technological system. A period of structural crisis in capitalist economies does not merely involve the Schumpeterian 'gales of creative destruction' in the economic sphere, but a restructuring of the entire sociopolitical infrastructure.

The challenge of institutional adaptation to the underlying process of technological and economic change resonates strongly with our concern with the fundamental questions of social learning. The extent to which a nation or region's capacity for technological learning and adaptation is supported or weakened by its institutional structure is critical to its success in a period of rapid economic transformation. The diverse approaches to the study of institutions outlined above present two apparently contradictory perspectives: one which views institutions as systems for organizing and constraining social behaviour through the reproduction and transmission of accepted norms and values; and another which views institutions more positively as mechanisms which can embed and preserve collective social knowledge about a wide range of subjects, especially those related to economic processes. In the first formulation, represented in the sociological tradition by Berger and Luckmann, or in the economic tradition by Veblen and Innis, institutions impede the process of social and economic adaptation. From this perspective, 'institutional rigidity' or 'institutional drag' is portrayed as a serious obstacle to change. The more optimistic perspective, such as that associated with Hodgson, views institutions as a vital mechanism for storing and transmitting the accumulated social knowledge critical to the process of change and adaptation. The key feature differentiating the two perspectives is the capacity for institutional adaptation and social learning.

A common thread running through both streams of the literature is the critical importance of institutions for incorporating social roles based on established norms and expected patterns of behaviour. Thus institutions represent a fundamental mechanism for socializing individuals into a wide range of roles across the numerous social fields where they interact, and they act as an important mechanism for transmitting information about accepted norms and expected patterns of behaviour. But the emphasis in this approach has been overwhelmingly on how individuals and firms learn, and it overlooks the fundamental challenge of how learning – and forgetting – occur within institutions themselves. Johnson (1992) suggests that in periods of rapid economic change the capacity for institutional forgetting may be just as important as the capacity for institutional learning. The accumulated inertia of existing habits or practices in economic and social institutions may block the potential for new learning processes. Old habits of thought and routines, even some norms and values, may have to be destroyed before existing social institutions can assimilate the new knowledge.

> The difficulties connected with creative forgetting constitute a risk for irrational lock-in of resources. Tax rules, capital markets, the character of competition and ownership and other institutional factors affect how these questions are handled (1992, p. 30).

This question clearly seems relevant for the much wider range of social and political institutions identified by Perez as well.

The more difficult question involves the issue of how institutions learn. Lundvall has been in the forefront of those stressing the need for an increased emphasis on learning in the new economy, arguing that it may be more appropriate to describe the emerging paradigm as a 'learning economy', rather than a 'knowledge-based' one. Learning in this respect refers to the building of new competencies and the acquisition of new skills, not just gaining access to information. The rapid pace of change associated with the 'frontiers' of economically relevant knowledge, means that its economic value tends to diminish the more widely it is disseminated. The easier and inexpensive access to information tends to reduce the economic value of more codified forms of knowledge and information. In tandem with this, forms of knowledge which cannot be codified and transmitted electronically (tacit knowledge) increase in value, along with the ability to acquire and assess both codified and tacit forms of knowledge, in other words, the ability to learn. In this sense, the dramatic effect of information and communication technologies on the rapid diffusion and availability of information and the emphasis on a 'learning economy' are integrally linked. It is the capability of individuals, firms, regions and nations to learn and adapt to rapidly changing economic circumstances that is more likely to determine their future economic success in the global economy (Lundvall and Borrás, 1998).

One way to approach the notion of institutional learning in the broader sense implied by Perez is through the concept of reflexivity. This idea is derived from a number of sources – not least the work of Anthony Giddens. For Giddens, reflexivity is grounded in the structures of social practice that are fundamental to his social analysis.

> Continuity of practices presumes reflexivity, but reflexivity in turn is possible only because of the continuity of practices that makes them distinctively 'the same' across time and space. 'Reflexivity' hence should be understood not merely as 'self-consciousness' but as the monitored character of the ongoing flow of social life (1984, p. 3).

Moreover he ascribes the characteristics of reflexivity not only to individuals but also to institutions. This element of *institutional reflexivity* has been picked up by Cooke (1997) who suggests that a capacity for self-monitoring must be viewed as an aspect of the institutionalized intelligence required to cope with the need for constant innovation that the industrial economies face in the context of continuous change and uncertainty. He suggests that the kind of institutional intelligence implied by the notion of *institutional reflexivity* implies a fourth level of learning, above and beyond the three referred to above – learning-by-learning – an essential element of the associational economy. This idea is elaborated further in Cooke and Morgan (1998), who see reflexivity as a crucial dimension of intelligence that is fundamental for the learning capacity of an organization or a region. They define reflexivity as 'the systematic process which combines learning and intelligence such that, in a number of feedback loops, the system receives guidance' (1998, p. 73).

Charles Sabel further develops this notion with his analysis of *learning by monitoring*. Sabel argues that the creation of discursive institutions where economic actors engage in discussion can play a critical role in reconciling the demands of *learning* with the demands of *monitoring*. By learning, he means acquiring the knowledge to make and do things valued in the marketplace; by monitoring, he means the ability of the parties involved to ensure that the respective gains from learning are distributed among them according to standards that they have agreed upon. The activity of discussion is critical for reconciling these two objectives, for 'discussion is precisely the process by which parties come to reinterpret themselves and their relation to each other by elaborating a common understanding of the world' (1994, p. 138; *see also* Helper, MacDuffie and Sabel, 2000).

Michael Storper (this volume) places equal emphasis on the importance of fostering public institutions that encourage concerned parties to commit to the kinds of conventions and relations that support an institutionalized learning economy. He sees talk as an essential process for generating these kinds of conventions and shared understandings. The value of talk arises from the need for communicative interaction that goes beyond the mere transfer of information between parties to build the conditions essential to achieve mutual understanding and acceptance:

> Talk refers to communicative interaction, designed not simply to transmit information and relay preferences, but to achieve mutual understanding. In the case of prospective learning, information from other experiences where learning has worked . . . can be valuable as a stimulus (p. 140).

The kind of talk that can build up this level of trust occurs most effectively within the context of public institutions, but the relation between talk and trust is highly circuitous – the inability to engage in the talk that can build trust and mutual understanding often reflects the absence of a tradition that values the presence of these kinds of public institutions. However, talk must be supported by a range of incentives that encourage the parties to maintain their involvement with these institutions. Small, repeated experimental interactions may prove effective as a mechanism for getting the parties to work together in a limited fashion and facilitate institutionalized learning.

Where this process succeeds, these institutions can play an important role in connecting the state to the economy, as well as various economic actors to each other. In institutions that foster learning by monitoring, actors can gauge the benefits they are gaining through their involvement without making themselves overly vulnerable. This allows them to achieve a limited degree of cooperation by defining common goals, yet continuing to scrutinize each other's actions. Sabel suggests that this process may be particularly beneficial in situations of rapid economic change and in the emerging knowledge-based economy, where the production of complex goods requires the coordination of many specialized firms across diverse branches of the industrial and service sectors. Where learning by monitoring has successfully been institutionalized in this way, it allows actors to assess reflexively where cooperation is advantageous and mutually beneficial, by placing responsibility for the process directly on the actors themselves (Sabel, 1994, p. 159).

The concepts of *reflexive learning* and *learning by monitoring* bear a certain affinity to ideas that have also been developed in the international relations literature. Ernst Haas has written about the importance of learning in international organizations. By learning he means 'the process by which consensual knowledge is used to specify causal relationships in new ways so that the result affects the content of public policy' (1990, p. 24). In this sense, learning occurs when members of the organization begin to question earlier beliefs about the appropriateness of the course of action that they are pursuing and to consider alternative ones – in other words, to re-evaluate their approach. The learning process involves the development of new common understandings of the problems that members of the organization face and consequently, a shared approach to the solutions. In this sense, learning implies a sharing of meanings among those who learn.

Both Ernst and Peter Haas employ the concept of epistemic communities to expand on this notion of how learning takes place in international organizations. According to the latter, epistemic communities (networks of knowledge-based experts) play an important role in examining the cause and effect relationships among complex problems, helping to frame issues for debate in international settings and proposing specific points for negotiation which may help to define or point the way to potential solutions or international agreements (Haas, 1992). He defines epistemic communities as 'a network of professionals with recognized expertise and competence in a particular domain or issue area' (1992, p. 3). Epistemic communities play an important role in helping overcome the problems of uncertainty and inertia that pervade international negotiations. Epistemic communities can help alter the negotiation strategies pursued by states in the international arena. States may respond to the new knowledge or new approaches generated by epistemic communities by pursuing new objectives or new negotiating strategies in the international context. Decision-makers are more likely to resort to the expertise available through epistemic communities under extreme conditions of uncertainty or following a shock to the international regime. Under these circumstances, epistemic communities provide insight by interpreting the cause and effect relationships that may have triggered the crisis or shock and they can help analyze the likely consequences of alternative courses of action. In situations of developing new or unprecedented approaches to dealing with international problems, epistemic communities can play a central role in defining new conceptual frameworks for dealing with the situation. Three of the essays included in this volume deal with just such processes: Robert Wolfe analyzes the process by which decision-makers have devised new conceptual frameworks to expand the regime governing international trade relations; Tony Porter documents how decision-makers in international bodies struggled to respond to shocks to the existing international financial regime; while Liora Salter examines a similar process with respect to international standards for dealing with the myriad of issues generated by new technologies.

What this literature also makes clear is that social learning dynamics have become more important recently at the international level, just as they have at the national and subnational (regional and local) scale, and for many of the same reasons. The emergence of new information and communication technologies has created the potential for internationally organized systems of production, trade and investment to be radically restructured. The same technologies have also created entirely new markets for telecommunications products and services, and these markets are themselves structured on a global scale. They have also enabled the liberalization of the international financial system, with the risk of substantially increased system-wide instability. Against this backdrop, learning plays a key role in enabling the institutions that regulate economic processes to evolve and adapt to change.

These concepts provide us with a useful way to approach the question of social learning and institutional adaptation posed above. The institutional capacity of individual regions, nations and international regulatory regimes to respond to the challenge of economic restructuring stimulated by the emergence of a new paradigm may largely depend on their capacity for *institutional reflexivity* – that is, their capacity to monitor reflexively their own ability to shed inefficient institutional norms and practices and replace them with ones that assist the process of economic change and social adaptation. The creation of new institutions, or the transformation of existing institutions at the regional, national and international level where this can occur, is an essential part of this process.

Outline of the book

The essays gathered together in this volume explore the challenge of how social learning occurs at the broader level of organizations and institutional frameworks. While we would not claim that they provide definitive answers to the questions posed thus far, collectively, they provide some valuable insights into a number of the questions that we have raised.

The expanding domain of the global market, both functionally and geographically, poses a major challenge to states wishing to maintain social control of the economy while promoting dynamic growth. Farms and phones may seem an unlikely pair for a discussion of international institutional innovation, but they represent two sources of enormous political controversy that dominated the Uruguay Round of multilateral trade negotiations between 1986 and 1993 – the old issue of agriculture, and the new issues of intellectual property, investment, and services as they relate to telecommunications. In his chapter, Robert Wolfe approaches the world trading system as a set of social institutions created by states for coordination of economic flows between states. He argues that the possibilities for continued successful adaptation of these institutions in the face of new challenges will depend first and foremost on how states learn about new problems and potential solutions. This learning phase precedes (and indeed, makes possible) bargaining and negotiation between states. In both telecommunications and agriculture, we cannot understand how the institutions of the trading system adapted to structural change during the Uruguay Round without understanding how states learned. States learned from other states, often with the support of epistemic communities; they learned from the requests made of them and from the offers made by others. Moreover, the most crucial forms of interaction took place in face-to-face meetings between state officials, in which talk and mutual persuasion led to the development of commonly held understandings ('consensual knowledge') of the most pressing trade issues of the day and potential avenues for their resolution. The greater role for global markets enabled by technological change has inevitably created a greater need for states to act together, thereby enhancing the importance of social learning on an international scale.

While it is commonplace to think of international finance as the domain of atomistic market actors, Tony Porter argues in his chapter that this sector of the world economy is in fact dependent upon a dense fabric of institutions, ranging from informal social institutions, such as the private tacit understandings that have regulated the Eurobond markets, to large formalized organizations such as the International Monetary Fund. Porter argues that the efficacy of these institutions depends crucially on their ability to evolve and adapt through learning, and his chapter analyzes the capacity of international financial regimes to learn. He argues that the system's capacity for higher level social learning has increased over time. But although we have seen a greatly increased capacity for communicating, producing, assessing and storing information, institutional fragmentation and the lack of open deliberation remain a serious problem. Hence, in contrast to Wolfe's analysis, Porter argues that in the realm of international finance, social learning leading to institutional innovation is reactive rather than proactive. In his view, instances of system-wide learning are sporadic and tend to appear most commonly in the wake of crises. This is most striking in the lack of coordination between the institutions concerned with prudential regulation and those concerned with the liberalization of financial flows.

Not all of the chapters presented in this volume are equally sanguine about the prospects for institutional learning and adaptation to occur in response to the emerging techno-economic paradigm. A critical precondition of the new information technology paradigm is the existence of standards for interconnection and for assuring harmonization. Liora Salter's chapter explores the processes and the institutional context of standard-setting in this paradigm. Standardization occurs in a complicated network of institutions, which are currently undergoing precisely the kinds of changes envisioned in this volume. New forms of organization are being created to take account of the particular demands of the emerging paradigm. However, Salter argues that institutions must meet two challenges to be worthy of the label 'learning institutions'. They must demonstrate creative adaptation within the new economy on its terms, and creative (that is, democratic and humane) responses to the pressures introduced by the new economy and its negative by-products. This case study of communication and information standards strains against much of the analysis presented in the rest of the volume in its negative assessment of the potential for collaboration, networking and social learning in these institutions. These findings must be factored into any discussion of 'learning institutions', and into policy prescriptions for adapting to the new economy.

Salter is not the only contributor who foresees potential impediments to social learning. Lynn Mytelka's chapter explores the implications for localized learning stemming from the turbulence created by globalized innovation-based competition and the heightened capital mobility. She is especially concerned with its effects on the learning environment for small and medium-sized enterprises (SMEs), many of which are suppliers within international production networks. Mytelka argues that inward foreign direct investment has the *potential* to complement and catalyze production locally and stimulate innovation through knowledge spillovers and the transfer of information and technology through customer–supplier linkages. However, it can also crowd out local competitors, strip proprietary knowledge and other assets from these firms through mergers and acquisitions, and engage in a variety of market-distorting practices with highly negative effects for the achievement of broader social and economic goals. Accelerated capital mobility, exacerbated at the regional level by what she calls locational tournaments, potentially erodes the basis for the development of localized learning economies.

Although the interest in social learning dynamics at the international and local or regional scales is well justified, it is important to acknowledge the enduring nature of national institutions and the key role they play in supporting these learning dynamics. While there seems to be growing consensus that shared, distinctive regional 'cultures' play a vital role in facilitating social learning processes leading to innovation, the origins of these cultures remain somewhat obscure. Much of the existing literature emphasizes the role of regional histories and institutions in shaping regional cultures of cooperation and facilitating the joint production and transfer of new product and process innovations. In his chapter, Meric Gertler argues that this conventional explanation may overemphasize the influence of regional institutions, while underemphasizing the importance of institutional forces at the scale of the nation state. When users and producers interact to generate new knowledge through social interaction, they share considerably more than spatial proximity and cultural or communicative commonalities. Through the use of a case study involving the transnational interaction between technology users and producers, Gertler demonstrates that effective social learning is underpinned by a shared set of rules, expectations, norms and practices that arise from a common macro-regulatory framework. Although firms may wish to collaborate, if their individual evolution has not been shaped by a similar set of national institutions, the likelihood of success in achieving effective interfirm learning will be considerably reduced.

Michael Storper reminds us that economies, to be successful, must be equipped to keep outrunning the powerful forces of standardization and imitation in the world economy. Once their firms' products are imitated or their outputs standardized, these economies are subject to downward wage and employment pressures. They must become moving targets by continuing to learn. Storper suggests ways to construct frameworks of action in the learning economy, and identifies four major steps in the formulation of such an economic strategy. The first is strategic assessment. The second step is the definition of the capacities for action and identities of actors, which are associated with the world(s) of production to be assisted by policy. The third step is the implementation of specific versions of heterodox meso-economic policies, whose content is defined by a combination of technical assessment and social process, especially talk. Finally, and only at the end of this long and 'soft' process, can the need for further formal institution-building be realistically assessed and practically undertaken, the latter on the basis of confidence-building precedent (and hopefully success in learning), and consequently emerging collective identities.

Despite continued predictions of the 'end of geography', regions are becoming more important modes of economic and technological organization in the age of global, knowledge-intensive capitalism. Richard Florida's central argument is that regions are themselves becoming focal points for knowledge-creation and learning, as they take on the characteristics of *learning regions*. In Florida's conception, learning regions function as collectors and repositories of knowledge and ideas, and provide an underlying environment or infrastructure that facilitates the flow of knowledge, ideas and learning. They have thus become increasingly important sources of innovation and economic growth within the globalized economy. He suggests that we are likely to see a shift from strategies and policies that emphasize competitiveness to ones which revolve around the concept of *sustainable advantage*. Sustainable advantage means that organizations, regions and nations shift their focus from short-run economic performance to recreating, maintaining and sustaining the conditions required to produce and support globally competitive firms over a longer period of time. According to Florida, the ability for government to adopt the principles of continuous knowledge mobilization and knowledge-intensive organization will become a source of sustainable advantage for firms, regions and nations in the twenty-first century.

Philip Cooke's chapter examines the extent to which regions can and do support innovation by firms and other organizations. He outlines the results of a large-scale research project on 'Regional Innovation Systems' funded by the European Union (EU) and judges the degree to which diverse European regions match up to the theory and practice of 'the new regionalism'. The research examines nine EU regions and two from Central and Eastern Europe in order to determine the extent to which the competitiveness of regions was related to their degree of systemic innovation capability. Three key points emerge from this account of innovation and competitiveness amongst firms in diverse parts of Europe. First, despite the hype about globalization, most European firms spend much of their time and energy operating mainly in regional and national markets. Second, firms innovate because to compete they must produce higher quality at lower cost. Finally, the organizations that exist to help firms innovate are failing to do so. They are not used and not respected by firms because they do not meet their needs or help them to identify their needs. Cooke concludes that the whole regional innovation and enterprise support system is in need of serious overhaul with re-focusing within the EU Framework and Structural Funding programs, and more innovative thinking and action on innovation from the regions.

One of the key questions in regional development today – a question which resonates for theory, policy and practice – is whether interactive learning networks evolve organically, through the repeated transactions of firms and their cognate associations, or whether they can be constructed through judicious public policy. Kevin Morgan and Dylan Henderson argue that a radically new kind of regional policy is emerging in the European Union in which the accent is on collective learning and institutional innovation rather than upon basic infrastructure provision. They also argue that these new regional innovation policies signal the most determined effort to date to build social capital, a relational infrastructure for collective action predicated on trust, reciprocity and the disposition to collaborate to achieve mutually beneficial ends. There is a danger that the indicators used to assess these policies may not be appropriate; for example, new regional policies, which aim to raise innovation capacity, might be judged by the standards of the old regional policies (short-term job creation) and be prematurely jettisoned because they fail to meet these standards. Regional innovation policy, as expressed in the RTP and RIS programs they describe, clearly has its limits. At present these programs are small-scale, low-budget experiments that have yet to be fully deployed even in the regions where they have been pioneered. To be effective, Morgan and Henderson argue, such programs need to be taken up and extended by national and supranational authorities in the EU, otherwise they might atrophy for lack of scale and resources. For all that, regional experimentalism might have some lessons for the higher echelons of the state, particularly as regards governance structures and policy-making processes, where politicians and officials too often think of themselves as tutors rather than learners.

Finally, David Wolfe argues that the process of social learning poses a particular challenge for older industrial regions with mature or established economies, such as those in the industrial heartland of North America and Europe. In these economies, institutional practices are embedded in well-established cultural and social practices. In some instances, these practices may support innovation and social learning, but in others, they may not be particularly well suited to the institutional requirements of the new paradigm of the learning economy. In these cases, the need to 'forget' may be a prior condition of the ability to learn. The inertial effect exerted by the power of old routines and habits may block the ability of firms or networks to develop new learning processes. Innovative attempts to stimulate a learning economy and create new associative forms of governance on a sectoral basis by the NDP government elected in Ontario in 1990 enjoyed limited success, but ultimately foundered on the resistance engendered by established cultural and social practices in some sectors and the political inability to forge a new development coalition to support the shift.

If all of the major studies of innovation agree on one thing, it is the finding that the most important innovations arise when previously separate and distinct bodies of knowledge are brought together in novel ways. It is our sincere hope that, by addressing ongoing conceptual debates in fields as diverse as economic geography, innovation systems, international relations, and the political economy of growth and development, the papers in this collection will themselves yield important insights into the structure of social learning dynamics and their role in the innovation process.

References

Barnes, T.J. (1999) 'Industrial Geography, Institutional Economics and Innis', in T.J. Barnes and M.S. Gertler (eds), *The New Industrial Geography: Regions, Regulation and Institutions*, London: Routledge.

Berger, P.L. and T. Luckmann (1967) *The Social Construction of Reality: a Treatise in the Sociology of Knowledge*, New York: Anchor Books.

Cooke, P. (1997) 'Institutional Reflexivity and the Rise of the Region State', in G. Benko and U. Strohmayer (eds), *Space and Social Theory: Interpreting Modernity and Post-Modernity*, Oxford: Blackwell.

Cooke, P. and K. Morgan (1998) *The Associational Economy: Firms, Regions, and Innovation*, Oxford: Oxford University Press.

DiMaggio, P. and W.W. Powell (1991) 'Introduction', in W.W. Powell and P. DiMaggio (eds), *The New Institutionalism in Organizational Analysis*, Chicago: University of Chicago Press.

Doremus, P., W. Keller, L. Pauly and S. Reich (1998) *The Myth of the Global Corporation*, Princeton: Princeton University Press.

Dosi, G., C. Freeman, R. Nelson, G. Silverberg and L. Soete (eds) (1988) *Technical Change and Economic Theory*, London and New York: Pinter Publishers.

Edquist, C. and B. Johnson (1997) 'Institutions and Organizations in Systems of Innovation', in C. Edquist (ed.), *Systems of Innovation: Technologies, Institutions and Organizations*, London and Washington: Pinter Publishers.

Freeman, C. and C. Perez (1988) 'Structural Crises of Adjustment, Business Cycles and Investment Behaviour', in G. Dosi, C. Freeman, R. Nelson, G. Silverberg and L. Soete (eds), *Technical Change and Economic Theory*, London and New York: Pinter Publishers.

Gerth, H. and C.W. Mills (1954) *Character and Social Structure: the psychology of social institution*, London: Routledge & Paul.

Gertler, M.S. (2001) 'Best Practice? Geography, Learning and the Institutional Limits to Strong Convergence', *Journal of Economic Geography* 1: 5–26.

Giddens, A. (1984) *The Constitution of Society: Outline of the Theory of Structuration*, Berkeley: University of California Press.

Haas, E.B. (1990) *When Knowledge is Power: Three Modes of Change in International Organizations*, Berkeley and Los Angeles: University of California Press.

Haas, P. (1992) 'Introduction: Epistemic Communities and International Policy Coordination', *International Organization* 46(1): 1–35.

Hall, P.A. (1986) *Governing the Economy: the Politics of State Intervention in Britain and France*, New York: Oxford University Press.

Hall, P.A. (1997) 'The Role of Interests, Institutions, and Ideas in the Comparative Political Economy of the Industrialized Nations', in M.I. Lichbach and A.S. Zuckerman (eds), *Comparative Politics: Rationality, Culture, Structure*, Cambridge, UK: Cambridge U.P.

Hall, P.A. and D. Soskice (eds) (2001) *Varieties of Capitalism*, Oxford: Oxford University Press.

Helper, S., J.P. MacDuffie and C. Sabel (2000) 'Pragmatic Collaborations: Advancing Knowledge while Controlling Opportunism', *Industrial and Corporate Change* 9: 443–88.

Hodgson, G.M. (1988) *Economics and Institutions: a Manifesto for a Modern Institutional Economics*, Cambridge, UK: Polity Press.

—— (1993) 'Evolution and Institutional Change: on the Nature of Selection in Biology and Economics', U. Mäki, B. a. K. Gustafsson and C. Knudsen (eds), in *Rationality, Institutions and Economic Methodology*, London and New York: Routledge.

Hollingsworth, J. Rogers (2000) 'Doing Institutional Analysis: Implications for the Study of Innovations', *Review of International Political Economy* 7:4 (Winter), pp. 595–644.

Innis, H.A. (1956) *Essays in Canadian Economic History*, M.Q. Innis (ed.), Toronto: University of Toronto Press.

Johnson, B. (1992) 'Institutional Learning', in B-Å. Lundvall (ed.), *National Systems of Innovation: towards a Theory of Innovation and Interactive Learning*, London: Pinter Publishers.

Johnson, B. and K. Nielsen (1998) 'Introduction: Institutions and Economic Change', in K. Nielsen and B. Johnson (eds), *Institutions and Economic Change: New Perspectives on Markets, Firms and Technology*, Cheltenham: Edward Elgar.

Keohane, R.O. (1990) 'Multilateralism: an Agenda for Research', *International Journal* 45 (Autumn).

Lundvall, B-Å. (1988) 'Innovation as an Interactive Process: From User–Producer Interaction to the National System of Innovation', in Dosi *et al.* (eds).

—— (1992) 'Introduction', in B.-Å. Lundvall (ed.), *National Systems of Innovation: towards a Theory of Innovation and Interactive Learning*, London: Pinter Publishers.

—— (1998) 'The Learning Economy: Challenges to Economic Theory and Policy', in K. Nielsen and B. Johnson (eds), *Institutions and Economic Change: New Perspectives on Markets, Firms and Technology*, Cheltenham: Edward Elgar.

Lundvall, B-Å. and S. Borrás (1998) *The Globalising Learning Economy: Implications for Innovation Policy*. Targeted Socio-Economic Research Studies, DG XII, Commission of the European Union, Luxembourg: Office for Official Publications of the European Communities.

Morgan, K. and C. Nauwelaers (1999) *Regional Innovation Strategies: Lessons for Less Favoured Regions*, London: The Stationery Office.

Mumford, L. (1934) *Technics and Civilization*, New York: Harcourt, Brace and World.

Nelson, R.R. (1988) 'Institutions Supporting Technical Change in the United States', in G. Dosi *et al.* (eds), *Technical Change and Economic Theory*, London and New York: Pinter Publishers.

—— (1994) 'The Co-Evolution of Technology, Industrial Structure and Supporting Institutions', *Industrial and Corporate Change* 3(1): 47–63.

North, D.C. (1990) *Institutions, Institutional Change and Economic Performance*, Cambridge: Cambridge University Press.

OECD (1994) *Accessing and Expanding the Science and Technology Knowledge Base*, DSTI/STP/TIP (94)4, Paris: Organisation for Economic Co-operation and Development.

OECD (2001) *Understanding the Digital Divide*, Paris: Organisation for Economic Co-operation and Development.

Perez, C. (1983) 'Structural Change and Assimilation of New Technologies in the Economic and Social Systems', *Futures* 15(5): 357–75.

Polanyi, K. (1957a) 'The Economy as Instituted Process', in K. Polanyi, C. Arensberg and H. Pearson (eds), *Trade and Market in the Early Empires*, Glencoe, IL: The Free Press.

—— (1957) [Inc.: 1944] *The Great Transformation*, Boston: Beacon Press.

Ruggie, J.G. (1982) 'International Regimes, Transactions and Change: Embedded Liberalism in the Postwar Economic Order', *International Organization* 36(2): 195–231.

Rutherford, M. (1994) *Institutions in Economics: the Old and the New Institutionalism*, Cambridge, UK: Cambridge University Press.

Sabel, C.F. (1994) 'Learning by Monitoring: the Institutions of Economic Development', in N.J. Smelser and R. Swedberg (eds), *The Handbook of Economic Sociology*, Princeton, NJ and New York: Princeton University Press and Russell Sage Foundation.

Schumpeter, J.A. (1950) *Capitalism, Socialism and Democracy*, Third ed., New York: Harper and Row.

Storper, M. (1997) *The Regional World*, New York: Guilford Press.

Tapscott, D. (1996) *The Digital Economy: Promise and Peril in the Age of Networked Intelligence*, New York: McGraw–Hill.

US Department of Commerce (2000) *Falling through the Net: toward Digital Inclusion*, Washington, DC.

Veblen, T. (1919) *The Place of Science in Modern Civilization and Other Essays*, New York: B.W. Huebsch.

Weber, M. (1978) *Economy and Society: an Outline of Interpretive Sociology*, Guenther Roth and Claus Wittich (eds), Berkeley: University of California Press.

Whitley, R. (1999) *Divergent Capitalisms: the Social Structuring and Change of Business Systems*, Oxford: Oxford University Press.

Williamson, O. (1985) *The Economic Institutions of Capitalism: Firms, Markets, Relational Contracting*, New York: The Free Press.

Wolfe, D. (1988) 'Socio-Political Contexts of Technological Change: Some Conceptual Models', in B. Elliott (ed.), *Technology and Social Process*, Edinburgh: University of Edinburgh Press.

Wolfe, D.A. (1994) 'State, Law and Economy: the Reflexive Basis of Institutions', Canadian Institute for Advanced Research, Program in Law and the Determinants of Social Ordering, Working Paper No. 8., Toronto, October.

Zysman, J. (1994) 'How Institutions Create Historically Rooted Trajectories of Growth', *Industrial and Corporate Change* 3(1): 243–83.

—— (1996) 'Institutions and Economic Development in Advanced Countries', in G. Dosi and F. Malerba (eds), *Organization and Strategy in the Evolution of Enterprise*, Basingstoke: Macmillan – now Palgrave Macmillan.

2
Farms, Phones and Learning in the Trade Regime

Robert Wolfe

Introduction

Farms and phones may seem an unlikely pair for a discussion of international institutional innovation, but they symbolize the two sources of enormous political controversy that dominated the Uruguay Round of multilateral trade negotiations (1986–93) – the old issue of agriculture, and the 'new issues' of intellectual property, investment and services. The theme of the Round was globalization, understood as change in the things that are traded, and in who is trading them. The expanding domain of the global market, both functionally and geographically, challenged states anxious to maintain social control of the economy while promoting dynamic growth, the basis of the 'compromise of embedded liberalism' underpinning the Bretton Woods system (Ruggie, 1983). The process of adaptation, of maintaining the compromise, proved to be similar in both domains.

My argument in this chapter is that the possibility of continued institutional adaptation, both organizationally and substantively, will depend first on how states learn about new problems, rather than on bargaining. Until states understand the issues, that is, until states can agree on the terms of debate – even on how to measure apparently rising interdependence in a given sector – bargaining is not possible. The world economy is not a concrete thing, nor does it exist in some state of nature: it is a social institution. States created the trading system, understood as the set of practices, understandings and rules that permit exchanges across borders. That system in turn has a powerful effect on how states regulate trade. The problem to be solved, and the means of solving it are both socially constructed within the institutions of the society of states. This can be illustrated by the way that international problems associated with farms and phones came to be seen as trade issues.

The trade ministers who assembled in Punta del Este, Uruguay in September 1986 agreed on the 'urgent need to bring more discipline and predictability to world agricultural trade.'[1] In contrast, a sense of crisis is absent from their call for negotiations 'to establish a multilateral framework of rules and principles for trade in services', though caution is evident in the stipulation that the 'framework shall respect the objectives of national laws and regulations applying to service. . .'. The very existence of the section was remarkable, given the determined opposition of some of the largest developing countries – Brazil and India led the fight against a new Round by claiming that the new issues had no place in GATT. To the surprise of some trade economists, however, it was agriculture that proved to be a bigger obstacle for the Round.

Life on the farm still follows ancient rhythms, but farmers are also affected by the practices of farm unions, government subsidies (at home and in far-off countries), the market machinations of seed companies, and the rules of such international organizations as the Food and Agriculture Organization (FAO) and the World Trade Organization (WTO). Similarly, the telephones in our homes, a service, are affected by: the state and its regulatory agencies; the rules of the WTO and the International Telecommunications Union (ITU); the internal operations of giant telecommunications firms; and the evolving practices of the Internet.

Both the old and the new issues involved the international problems associated with capitalist expansion. Changes in the pattern and volume of trade, for farmers as for phone calls, created conflicts as new participants entered the trading system and established traders experienced international pressures in unexpected ways. The task of the WTO, as with any economic international organization, is to *foster industry* by creating and securing international markets for industrial goods, and 'to *manage potential conflicts* with organized social forces which might oppose the further extension of the industrial system promoted by the activities undertaken to complete the first task' (Murphy, 1994, pp. 34, 192, 262). Fostering industry, which is easier than managing social conflict, more closely describes the task of negotiations on telecommunications services. The larger subject of the agriculture negotiations, in contrast, was managing the social conflict manifest in the Farm War.

The issues are comparable in another way: neither was negotiable at the outset of the Round. States do not always know what they want. Problems do not emerge fully formed on to a well-defined international agenda. How do governments know they have a *public* 'problem' susceptible to solution by 'policy'? How do they identify potential solutions? The thousands and thousands of international meetings that take place every year are not evidence that bureaucrats like to travel. They are instead evidence that state officials talk to each other in attempts to persuade, that through social interaction their understanding of their interests can change (Finnemore, 1996, pp. 121, 141). It was not obvious at the outset that the sense of crisis felt by temperate zone grain farmers would require both cuts in 'export subsidies' by some countries and reductions in aggregate domestic support by others. Nor was it obvious that the convergence between computers and telephones would lead to a recognition that telecommunications was a tradeable service.

Many constructivists are interested in where new norms come from, but we should also be interested in how existing norms construct new situations.[2] Following Ruggie, in situations where 'shared social purposes' endure, we should expect to see the normative framework of international regimes playing an important role. Structural change creates new commodities, undermining existing regulation. The process of reconstructing regulation is then shaped by the norms of the regime. In this chapter I try to show that this normative process is central to the way domestic regulation responds to the expansion of traded services, and that the norms of the trade regime ensure that the outcome is a process of ongoing intersubjective accommodation rather than a definitive harmonization of interests.

In the next section, I discuss what I mean by a 'regime'. I then offer a brief discussion of how agriculture fared in the Uruguay Round. The largest section of the chapter is an extended discussion of how states learned about telecoms. I show that learning precedes bargaining, that evolving regulatory practice and the international regime are mutually constitutive.

Regimes and learning

The internationalization of regulation is not a new story. Each era of industrial change brings new needs. 'International regimes' arise when underlying transaction flows representing an increase in the role of markets meet a social response from governments in what Polanyi (1944) called the double movement. We see regimes first in shipping (for example on the Danube), telegraphy (the ITU), railways (standard gauge in Europe), and trade (the GATT although drafted in 1947 codified a regime already at least a century old). This classic role for international coordination arises when firms, often in response to new technologies, find themselves operating in markets that have outgrown regulatory boundaries. The multilateral trading system, the 'trade regime', is a social institution created by governments to manage their interdependence with each other and to manage the interdependence among their trade-related policies.

The trade regime is made concrete in the WTO. The WTO is a structure of rules elaborated on the basis of the normative framework of the trade regime, and it is a container for a set of commitments among countries that reflect the specific application of those rules.[3] The GATT and now the WTO is not and was never meant to be a 'free trade agreement' aimed at the elimination of all barriers to commercial exchange. Economists believe that free trade is the best policy, even if states often think or act otherwise, but they do not agree on the details of what 'free trade' means in every circumstance in an era of rapid economic change, and their views evolve. Even if economists did agree, consistently, expertise in itself never makes a policy politically legitimate. Part of what is essential in making trade policy – as in any other domain – is the means available to governments to learn about and agree on appropriate policy.

The trade regime, therefore, reflects a given constellation of economic power and legitimate social purpose. *Structural* change in the world economy might be expected to lead to pressures for *instrumental* change within the regime, but if the social purposes of states are constant, we would expect those pressures to result in change within the existing normative framework (Ruggie, 1983). Where the norms of the GATT have proven remarkably robust, the instrumental form of the regime has been in constant evolution, as we might expect given the considerable shifts in the structure of the world economy in recent decades. When the domain of the market expands, as it has with globalization, the trade regime also expands. In situations of rapid change, we would pay attention to: how regimes can provide transparency, helping states understand each other; how they can legitimate new courses of action; and how they can be the forum for political contention about what is or is not 'knowledge' about a problem (Kratochwil and Ruggie, 1986, p. 773).

Liberalization, therefore, is a political process. It does not cause economic change, but it ensures that government policy accommodates change. The WTO seeks to manage conflict between states while respecting national differences. The WTO does not regulate transaction flows, but it does attempt to reconcile how *states* may regulate those flows. That is, the trade regime is a place for the adjustment of states' reciprocal or convergent expectations about appropriate behaviour. States shape and are shaped by the characteristic way in which the WTO, or the trade regime, seizes an issue. Over time, the way the WTO thinks can change instrumentally but not normatively. It will always ask if and why a thing is a trade issue. Only borders give the WTO a role: it cannot conceive of a problem that does not involve sovereign governments. Governments, not firms, are the *subjects* of its normative order, even if the topics under discussion involve people and firms as well as goods and services. The WTO is charged with finding ways to *reconcile* the differences brought into high relief by interdependence, rather than with *removing* difference. How states understand themselves in relation to the world trading system – their *identity* as traders – has been shaped by the trade regime just as states shaped the regime, as we can see in the response to the Farm War.

Farm trade as an international problem

The Farm War was a major crisis for the trade regime and for farmers. The war was evident in disruptions on the farm and in world markets, in conflict among major governments, and in disagreements in international organizations. Farmers are now a small component of the labor force in the advanced economies, and agriculture's share in world trade has shrunk from 46 per cent in 1950 to barely 11 per cent now. During the same period, however, food production kept growing, and the importance of global markets for producers in OECD countries increased significantly. Foreign markets could not absorb all the food produced however. The international conflict I call the Farm War erupted in the early 1980s when governments responded to a rising surplus by rapidly increasing subsidies.[4] One part of ending the war was institutional adaptation in the trade regime.

Farm policy in all countries, industrial and developing, is made for domestic reasons. The domestic objective is sometimes food security but it is usually political stability. The social purpose of managing the process of structural adjustment generally means that the majority must make concessions to the minority. In industrial countries, city dwellers being a majority help the tiny minority of the population who are still farmers. In developing countries, unfortunately, the situation is reversed. Farm policy crosses the border when people in other countries are expected to bear the costs of domestic adjustment. In industrial countries, the locus of the Farm War, this externalization happens in one of three stylized ways. First is simple protection: foreigners are denied access to our markets, ensuring that home-grown food will have pride of place on our tables. Second, we give our farmers subsidized help in selling what they produce in some other market, somewhere. Third, we help our farmers directly. Since farmers do not like explicit charity, we pay subsidies only for what they produce; production then rises for reasons unrelated to effective demand. In all of these cases, the burden of adjustment is actually shared between foreigners and domestic society. One of the ways in which trade policy and farm policy are linked, therefore, is in distributive conflict over where the burden will rest most heavily. During the Farm War, adjustment costs in the industrial countries were externalized to world markets, where price volatility was an effect not a cause of protectionism. These adjustment costs hurt food exporters in all developing countries, and they hurt food producers in importing countries. It is this process of externalization, the way in which farm policy affects commercial exchanges across borders, that has led to the international organization of agriculture being functionally part of the trade regime.

The Uruguay Round Agreement on Agriculture is an addition to the GATT 1994, the provisions of which still apply to farm trade.[5] It implicitly and explicitly incorporates, therefore, the norms and principles of the GATT. In common with other agreements, it contains a set of general rules and 'schedules' in which Members list their specific commitments. The Agreement was designed to end the Farm War by opening markets and reducing subsidies; it was also designed to prevent a recurrence of the war by providing international rules to guide the evolution of national policy while allowing room for those domestic policies to reflect differing national priorities. The two principles are that policy should support gradual adjustment in this sector and that support should be given in the least trade-distorting manner possible.

The implication of these principles is that countries can use whatever mix of policies seems appropriate in their own circumstances, so long as they attempt to minimize the externalization of the effects. The Agreement is above all a framework in which to capture and make commensurable the three different ways in which all countries help their farmers.

First, under market access the process of 'tariffication' converts all measures that affect access to a country's markets into bound tariffs subject to reduction. Second, all measures that provide domestic support for farmers are captured in one of the so-called boxes – Green, Amber or Blue, corresponding by analogy to traffic lights to whether or not the measures are subject to reduction. The Green Box was the most important, since it sets out the terms under which certain domestic policies that help farmers without affecting production or trade will be beyond the reach of WTO rules. Third, all export subsidies that help farmers sell in some other country's markets are captured in the Red Box and subject to reductions over the transition period. The creation of this framework can be understood as the rules negotiation. States also negotiated on 'Modalities', the method and the magnitude of reductions in measures captured within the framework. It is understood that this Agreement is only a cease-fire, because the reductions in the end will not be that significant. The 'continuation clause' provides, therefore, for further negotiations after a transitional period, and the 'peace clause' attempts both to limit disputes during the transition while encouraging the resumption of negotiations. A process for mutual surveillance of the process of implementation and subsequent review of the operation of the Agreement is also included through the creation of a new Committee for Agriculture, which reports to the WTO Council for Trade in Goods.

The new rules are much more significant than the limited liberalization achieved. The first objective of the round was ending the war, and the second was creating the basis for future stability. These objectives required three things: a reduction of export subsidies, new rules, and a mechanism for future negotiations. Liberalization was an objective only to the extent that it was a necessary tool of the central objectives. States agreed to no more liberalization than necessary. During the 1980s, changes in global food prices and in the food surplus were an important material force affecting world politics, but the social objectives of farm policy were constant. States saw the policy implication of the Farm War as a reconsideration of policy tools, rather than a reconsideration of their policy objectives or the creation of a new international regime.

The framework is a major conceptual and political achievement because of the comprehensive way in which it deals with border measures, subsidies and domestic policy. This framework evolved slowly through many years of informal discussion as well as negotiations in the Uruguay Round. The first stage of the negotiations was finding the objective. This creative process precedes bargaining, it is the process of learning what to bargain about. The process began in the early 1980s with efforts at the OECD to understand the problem; it ended with detailed bargaining on market access and with the creation of new surveillance and dispute processes to monitor compliance with the new rules. In the first stages, before the launch of negotiations in 1986, we see efforts to define the problem by deciding what things should be counted, and how. The political challenge was finding a consensus on the extent to which farm policy affected trade. In the middle stages, before and after the Montreal Ministerial of 1988, we see efforts to construct a framework in which to make commensurable the things being counted. The later stages, after the Brussels Ministerial of 1990 and the draft Final Act of 1991, involved bargaining about how much to reduce the things that could now be quantified. The bargaining certainly reflected how states understood their interests, but the bargaining was only possible because that understanding changed during the 1980s.

The new rules are technically and politically sophisticated. In agriculture, border measures and export subsidies merely support domestic policy. Without including all policies, even minimal reform would have been impossible. The biggest conceptual challenge was finding a way to make all of agriculture policy negotiable. Tariffs are easy, but how does change in Canada's 'Crow Rate' transportation subsidies equal change in the EU variable levies that insulated the EU Common Agricultural Policy (CAP) from world markets, or a reduction in the volume in American subsidized exports of wheat? An important step was agreement that certain kinds of 'non-tariff' barriers could be turned into a tariff with equivalent protective effect. The other was the development at the OECD of the Producer Subsidy Equivalent (PSE) and the Consumer Subsidy Equivalent (CSE). The OECD defined a PSE as 'the payment that would be required to compensate farmers for the loss of income resulting from the removal of a given policy measure' (OECD, 1987: 100). The PSE was not itself a negotiating target in the Agreement – it is a measure of income transfer not trade distortion, making it an analytic not a policy tool. The PSE is useful if the objective is the reform of *agricultural policy* not the reform of *trade policy* – it measures the effects of all agricultural policies that may distort national resource allocation without necessarily distorting international trade. The PSE helped states grasp the extent of the problems, and it helped states understand how the trade effects of domestic policy could be aggregated for negotiating purposes, a central conceptual innovation of the Round.

Every participant in the agriculture negotiations, and especially the EU and the USA, tried with some success to ensure that the shape and details of the Agreement accommodated their own approach to farm policy. But even 'interest' is shaped by the regime, as are feasible bargains – the very process of interaction (for example, in negotiations) can be transformative of 'identity' and 'interests'. The negotiations first framed the problem to be solved and agreed on objectives. The disruption of the war was obvious, but the sources and dimensions of the war were not. It took time for states to create a comprehensive frame in which domestic policy, border measures and export subsidies could all be seen to be part of one 'trade' problem. During the first part of the process, states also developed new concepts, including the idea of aggregation. In the second part, they created new instruments. The very public process of lengthy negotiations served to signal the direction of change in policy to farm groups, agribusiness and national policy-makers, who immediately began accommodating themselves to the ideas in play. The nature of those changes in practice, in turn, influenced the ability of negotiators to come to an agreement. The result is that the Agreement legitimized policy change that was already underway, rather than itself being a force for change. This process allowed the adjustment to begin long before the negotiations were completed; indeed, it was arguably that adjustment that allowed the negotiations to conclude. In some sense the Final Act simply records that change in practice and binds it against backsliding. The WTO does not lead the process of change. The Agreement simply confirms that everybody will be doing the same thing, it locks in those changes, and creates stable expectations. The same processes through which the trade regime shaped the way states learned how to measure the effects of change and through which an agreement locked in change in state practice, can be seen in the domain of telecommunications.

Trading services

A significant indicator of the ability of institutions to adapt to change was the creation of the General Agreement on Trade in Services (GATS) in the WTO. This adaptation was a response to structural change, but it was dependent on a change in ideas. During the period from 1985 to 1995, trade in goods increased two and a half times, but trade in services increased over three times, reaching nearly US$1.2 trillion, which represented more than 20 per cent of world trade in 1995 (WTO, 1996, p. 23).[6] Trade in goods was over US$5 trillion that year, but trade in services is now more important than agriculture and mining together. The growth in the importance of services to the advanced economies is not a recent development, but the idea that services could be traded is. Until the mid-1980s, such trade could hardly even be measured.

The idea of 'trade in services' was first articulated in a 1972 report by a group of trade experts at the OECD, and many of the concepts were elaborated over the ensuing two decades in the Trade Committee. Adding the topic to the declaration launching the Uruguay Round in 1986 was difficult because many countries simply did not understand the concept or how it could be, or should be, negotiable. Drake and Nicolaïdis show how an 'epistemic community' of experts helped states understand the issues, arguing that *ideas* mattered more than interests until the issue had crystallized –

without consensual knowledge, the USA (major instigators of the 'new issues') could not force others to negotiate. Negotiations were conceptual for years until governments learned what to do. The influence of the epistemic community declined when states understood the material interests at stake in the bargaining. Using the GATT as a forum for negotiations on services was not obvious. The USA could have pursued bilateral deals, and liberalization might have happened on its own. Drake and Nicolaïdis conclude that

> GATT and its procedural framework became unavoidable because of the plausibility and spread of causal beliefs that services transactions had common trade properties, faced common trade barriers, and could be governed according to common trade principles (Drake and Nicolaïdis, 1992, pp. 51–2 , 98–9).

Unlike in goods, where, for example, 'agriculture' is only defined by reference to the chapter headings of the Harmonized System of Tariffs, governments had first to define the domain of GATS, which took many years. The agreement defines services (Article 1) by the four ways or 'modes' in which they can be supplied. Two are relevant for telecommunications: 1) services supplied from one country to another (for example, international telephone calls), officially known as 'cross-border supply'; and 2) a foreign company setting up subsidiaries or branches to provide services in another country (for example, a foreign cellular operator setting up operations in a country), officially 'commercial presence'. Once services were understood to be tradeable, the trade regime was constitutive of the new rules. GATS is no more about 'free' trade than GATT. It creates a structure of rules placing some constraints on how governments regulate, but leaving them free to pursue national policy objectives. The bargain is still the embedded liberalism compromise: liberalize trade while maintaining national regulation, so long as foreigners are treated fairly. Much of the negotiation concerned learning what the most-favored nation (MFN), national treatment, and transparency norms might mean when applied to services. And states had to learn what these principles might mean sector by sector, beginning with telecommunications.

In February 1997, Members of the WTO concluded a major negotiation on trade in basic telecommunications services.[7] States had been unable to agree on how to promote trade liberalization in this sector during the Uruguay Round, and they missed earlier deadlines for concluding an agreement. This positive outcome in international regulation was hailed as an historic achievement,[8] though international cooperation on telecommunications goes back to the creation of the ITU in 1865. The telecommunications regime (Cowhey, 1990; Zacher and Sutton 1996) centered on the ITU has proved so robust that trade in phone calls only moved into the ambit of the trade regime in the 1990s. As we might expect from the discussion of regimes above, structural change created a need to adapt the rules applying to telecommunications. Telecommunications could only move to the trade regime, however, when two conceptual changes happened. One was the international consensus that it was possible to trade services. The other was the idea that telecommunications was one of the services that could be traded.

The telecommunications sector has a dual role: it is a distinct sector of economic activity; and it is an underlying means of supplying other economic activities. Telecommunications firms are enormous in their own right, and they provide the infrastructure of modern trade routes on which other industries depend, from airline reservation systems to electronic money transfers. Global telecommunications trade in 1996 was $115 billion, up from $50 billion in 1990, with the larger portion representing equipment sales. Global telecommunications services were worth more than $600 billion in 1995, of which international services accounted for nearly $53 billion. Traffic is measured in 'settlement minutes', the accounting convention established by the ITU under which countries share the revenue from an international call. Global settlement reached 60 billion minutes in 1995, up from 4 billion only twenty years earlier (*see* Tarjanne, 1997; *The Economist*, 1997, p. 119) This increased importance of foreign markets was driven by technology, globalization, and changes in the commercial structure of the industry. The technological story is too complex and too often told to need repetition here. The main elements are convergence between computing and communications, development of fibre optic networks, and the growing use of satellites, with the result that the marginal cost of providing ubiquitous global high-speed communications approaches zero.[9] Globalization of economic activity has been facilitated by the transformation of communications, but the expanded geographic scope of business has also created a demand for a low-cost global communications infrastructure. Finally, the commercial structure of the industry based on large monopolies has been transformed, especially by privatization: since 1984, dozens of public telecommunications operators have been sold.

The telecommunications regime used to be a coordination game between national monopolies, ensuring interconnection and an allocation of property rights over the electromagnetic spectrum. The regime, with a set of readily observable principles and norms, consists of a complex web of international bodies, mostly part of the ITU, that deal with jurisdiction, damage control, technical barriers, and prices/market share. When markets in telecommunications were based on territory, the regime worked. It began to break down when firms became able to operate in markets bigger than states, and when technological change undermined its commercial assumptions. Basic telecommunications is just one part of a complex regulatory domain in which technological change blurs the distinction between 'international' and 'domestic' because interconnection between the networks of different firms takes place everywhere rather than being confined to gateways (Zacher and Sutton 1996, pp. 130–1, 136–9, 157). Nevertheless, it is still true that 'International telecommunications take place overwhelmingly through the interconnection and interoperability of nationally regulated telecommunication systems, not through some over-arching global network' (Woodrow, 1991, p. 325). As technology changed the sector, therefore, the challenge was cooperation among states not drafting global regulations. And the challenges were mostly not technological – the growth in settlement minutes testifies to the technical success of the ITU – but commercial. Successful technological adaptation had undermined the institutional structure of the industry.

Negotiators had to find ways to have GATS cover telecommunications as a sector in itself and as a trade route for others. The latter was simpler than the former. The GATS Annex on Telecommunications contains core obligations on access to and use of 'public telecommunications transport networks and services'.[10] It requires each Member to ensure that all service suppliers seeking to take advantage of scheduled commitments are accorded access to and use of public basic telecommunications on a reasonable and non-discriminatory basis. The Annex addresses access to these services by users (for example, airline reservation systems) rather than the ability of suppliers (for example, phone companies) to sell such services. (Note that anything to do with broadcasting, which can be politically and/or culturally sensitive, is not covered.) Governments did not offer commitments on basic telecommunications during the Round, but specific commitments on value-added telecommunications services were included in many schedules.

At the end of the round, therefore, the GATS applied to telecommunications services necessary for the provision of other services – essentially access to such telecommunications services needed, for example, by financial services firms – and to value-added services. Few commitments had been made on basic voice and data transmission, the largest part of the market, and some large Members were proposing to take MFN exemptions on these basic services. A new Negotiating Group on Basic Telecommunications was created in an effort to avoid locking-in these derogations from the new rules. Participants in the group agreed to negotiate on all telecommunications services both public and private, including voice telephony, data transmission, telex, telegraph, facsimile, private leased circuit services (for example, the sale or lease of transmission capacity), fixed and mobile satellite systems and services, cellular telephony, mobile data services, paging, and personal communications systems.

Understanding market access in this sector posed two special problems. First, in negotiations about trade in goods, market access can be defined according to relatively narrow tariff lines, and negotiations can be conducted bilaterally between the parties having the largest import interest in the given commodity in a given market. The resulting agreement can be extended to all other Members with relatively little pain under the MFN rule because the bargain was acceptable to the largest participants. Bilateral negotiations certainly were essential in creating the telecoms package, but they were insufficient. As in services generally, the barrier to entry is not a tariff at the border but a domestic regulation. Removal of a regulation along with 'national treatment' can be tantamount to an instant zero tariff, which MFN would then extend to all WTO Members. Once a barrier is reduced sufficiently to allow one foreign entrant, MFN would allow all foreign entrants, yet only one of them would have made a reciprocal commitment. The second special problem arises when domestic regulation enforces a monopoly, since national treatment can then amount to no access at all. Allowing a foreign service provider 'access' to the market is meaningless if rules on pricing allow cross-subsidization by a current or former monopolist.

In response to both problems, the USA added in a footnote that its 1995 offer on basic telecommunications was

> contingent on the agreement by a critical mass of WTO members to provide market access and national treatment for basic telecommunications services, as well as to provide commitments similar to those offered by the United States on pro-competitive regulatory disciplines (*Financial Times*, 1995, p. 5).

Critical mass meant two things. It required substantial offers from the largest markets, and offers comparable in form from countries that are relatively small market participants, since market access for one is not possible without market access for all. In the event, 'critical mass' in basic telecommunications meant first, commitments from 69 governments (contained in 55 schedules, the EU-15 counting as one). All industrialized countries participated, along with developing and transition countries from virtually every region of the world. The markets of the participants were said to account for more than 91 per cent of global telecommunications revenues in 1995, though the offers are so complex that the extent of liberalization is a matter for experts. And critical mass meant, second, acceptance of common principles on domestic regulation that would ensure that the access was real.

The telecommunications deal was presented in the press as the outcome of a big bargain associated with 'liberalization'. The press reported, rightly, that the role of the USA was critical, not surprising given its centrality. Of the top 30 international routes in 1995, five of the top six involved the USA (the two biggest routes being with Canada and with Mexico), and all but two involve either the USA or an EU member, or both.[11] Yet the American 'interest' in its 'pro-competitive model' owes much to its self-perception as a liberal state, just as the basic/enhanced distinction owes much to identity and nothing to interest. If another state had had US predominance, they might have seen their interests in other terms.[12] And when we look more closely, we see that the deal was also the outcome of a complex process of learning how to regulate and what to bargain over.

In the late 1980s, telecoms people thought they knew very little about international 'trade' in telecommunications services (Langdale, 1989, p. 220). But it was already clear from a survey (in which officials were asked about their ideas) that large numbers of trade experts and telecoms regulators saw new forms of modern telecommunications as being tradeable services (Woodrow, 1991, p. 331). Prior to that time, a great deal of learning about telecoms services had been going on at the OECD, as it had been for farm trade. Officials had been looking at the linkages for more than a decade in the Information, Computer and Communication Policy Committee's Working Party on Transborder Data Flows. In 1988, a new Working Party on Telecommunications and Information Services Policies began work in conjunction with the Trade Committee work on trade in services. At the OECD, officials were able to explore concepts, especially the applicability of ideas about trade and regulation. They discovered a certain similarity in the broad objectives of telecommunications policy, and differences in regulatory frameworks due both to industry structure (monopolies pose different problems from competitive firms) and national characteristics. In its work we see discussion of ideas that later crystallized in WTO, such as how to transfer 'national treatment' and MFN from OECD instruments and the GATT context to telecoms (OECD, 1990, pp. 9, 13, 29–30).

The process of learning was especially important for developing countries, who often had no clear view on these questions. They will need help in creating a new regulatory structure; they may get it from the ITU. The ITU was initially cautious about the idea of trade in services, but Pekka Tarjane as Secretary-General since 1989 has been working to transform how regulators see themselves. He sponsored a series of meetings to help the ITU community (regulators and business) understand what trade in services meant for their sector.[13] This explicitly intersubjective process engaged domestic regulators with trade experts in learning regime concepts.

All observers agree that one of the strongest factors influencing international change was the change in regulatory strategy first seen in the United States in the so-called 'deregulation' of the sector in 1984. The new approach encourages competition as a means of ensuring adequate provision of public services at an affordable price. Previously, regulators had assumed that since telecommunications was a 'natural monopoly', universal access to a telephone could only be ensured by specifying the price and quality of services provided. New ideas on regulation make the idea seem outmoded. American demands for 'market access' implied a demand for regulatory reform in other countries that would be consistent with the American 'procompetitive' model. (Arguably the initial creation of the GATT was part of an American attempt to externalize their domestic regulatory structure, in effect making the world safe for the New Deal, so their approach to farm trade and telecommunications was not new. The Blue Box is designed to accommodate the structure of American and EU farm support programs (*see* Wolfe, 1998) and the telecommunications agreements are based on the American distinction between so-called basic and enhanced or value-added services.)[14] The telecommunications deal included another first, in that the regulatory principles to which participants subscribed were the first WTO provisions on domestic competition policy. Negotiating trade in basic telecommunications services, therefore, was of necessity a negotiation about regulation of telecommunications. The issues then involved competition and investment policy as much as trade policy. The deal required investment (foreign ownership) and competition policy provisions, because foreign firms needed assurances that regulation would be fair and even-handed, and that former national monopolies would not be able to abuse their once dominant position. Negotiators decided that these principles should be made a part of the GATS subject to dispute settlement as a way of safeguarding the value of the market access commitments. Principles covering competition safeguards, interconnection guarantees, transparent licensing processes, and the independence of regulators are included in a text called the Reference Paper (RP). Most participants included the RP under 'additional commitments' in their schedule.[15]

The RP will shape how states reregulate in this domain. That is not a typo. Regulatory reform is often mistakenly seen as leading only to 'deregulation' (Cerny, 1991), but it can be understood differently. In Polanyi's terms, an increase in the role of the market will tend to destroy some existing forms of social organization. The usual response is to use the state to create new social forms, which work to regulate the market in new ways to achieve enduring collective purposes. For example, Vogel found in his case studies of the telecommunications sector that after competition is introduced new rules are needed to constrain an established operator from using cheap calls to drive out small re-sellers by 'rebalancing' local and long distance rates (Vogel, 1996, p. 30). Deregulation is a misnomer. Most often, the introduction of more competition (liberalization) is accompanied not by the reduction of government regulations but by reregulation – the reformulation of old rules and the creation of new ones. The result is 'freer markets and more rules' (p. 3). These new rules must be seen in the double context of efforts by the trading system to accommodate domestic regulation, and efforts by society to accommodate the trading system. Domestic regulation is reconstructed in the trade regime.

In shaping reregulation, or an ongoing process of learning, the RP was designed to shape: 1) regulatory institutions: for example, whether the regulator is independent of the incumbent telecommunications operator and national industrial interests; 2) regulatory processes: for example, whether there are measures ensuring that the decision-making process is known, and is non-discriminatory; and, 3) substantive regulatory policies: for example, policies concerning interconnection between carriers. The RP seems remarkably imprecise, to the point where it can only have meaning in the intersubjective interpretations of the participants. The regulatory principles of the RP, in short, seem to allow enormous national latitude in practice. The requirement is one of regulatory rhythm not harmony.[16]

Assessments of why the agreement proved possible in the end noted that countries and firms had learned. Developing countries saw they could lose from closure and benefit from openness. They need a modern infrastructure, it will be hugely expensive, and rules on foreign investment and regulatory transparency give business the confidence to invest. Modern telecommunications is an essential input to many other economic activities, but states risk not having access to the latest technology if firms think the costs of providing the service would be too high. Firms saw that losses from liberalization in their own markets could be made up by new opportunities abroad. Firms also benefit from the agreement because it is much easier to negotiate market access rules once, in a multilateral forum, rather than incurring the high transactions cost of bilateral access agreements. Still, the telecommunications agreement trails what is actually going on in the global market – it will not itself be the major force for liberalization. Change in the WTO may be consistent with the change desired by the USA, but policy in other states was also changing. In Canada, for example, telecommunications regulation changed in the 1990s for domestic reasons, albeit taking account of international pressure.[17] Despite intense pressure from the USA, Canada relaxed its rules on investment in this domain, but still will not allow majority foreign ownership. Similarly, the 1995 European Union offer on market access effectively did no more than offer to extend internationally, and bind, the internal liberalization then planned for 1 January 1998 (*Financial Times*, 1995, p. 9).[18] Often states only agree internationally to changes well underway domestically. The new order for phones, therefore, has important similarities to the new order for farms. The negotiations were as much about learning how to regulate in an era of structural change as they were about bargaining; states adopted the new regulatory mode in practice in advance of agreeing to written rules. The new order still attempts to reconcile multilateral openness with national distinctiveness in regulatory means and objectives.

Conclusion

The agreement on trade in basic telecommunications services concluded a period of intense bargaining about how members of the WTO would give each other access to their formerly closed markets for phone calls. It was also a moment in a process, now going back over 25 years, of states changing their understanding of their interests through discussion in international regimes. Similarly, when the Uruguay Round began in the mid-1980s, states were at war over farm trade. The agreement ending that war depended on bargaining over who could sell how many tonnes of wheat, but it also depended on a new understanding of the relation between farmers and international markets. In both domains, the bargaining is not without interest, but we cannot understand how the institutions of the trading system adapted to structural change without understanding how states learned. States learned from other states; they learned from the requests made of them and from the offers made by others. Twenty years ago, statistics on 'telecommunications services' did not exist; now, remarkably detailed commitments for trade in such services have been made. In the early 1980s the GATT dispute settlement system was close to collapse because Europeans and North Americans could not agree on the meaning of 'subsidy' as applied to agricultural trade. What we observe is a debate about appropriate regulatory structure and its role in relation to the market. States are constitutive of international regimes, but in both these cases, the regime will be constitutive of the state too.

The final question is explaining why farms and phones are trade issues. Why was the trade regime the locus of action rather than international organizations concerned with agriculture and telecommunications? Both the ITU and the FAO serve certain purposes for states. In the case of the FAO, its purposes concern food and farmers, not trade. Similarly, the ITU collects information, provides a forum for discussion of definitional or property rights issues, and helps telecommunications people understand their problems, but it has no trade norms and no mechanism for distributive bargaining. The ITU is an organization that helps states pursue their common ends (making the telecommunications system work) but it is ill-equipped to facilitate distributive bargains. An increase in (perceived) interdependence can create the need for more reciprocal agreements, and more explicit agreements, which creates a need for organizations that can manage distributive interaction. The ITU worked well for over a century, but when technological change ended the separation of markets forcing states to consider sharing their own and far-away markets, the ITU could not help. Policy had limited the boundaries of markets, in both farming and telecommunications, but technological and commercial change put great pressure on these social boundaries. But the role of policy in pursuing national objectives is still seen as legitimate by states. The regulatory task has changed, but not its objective. The greater role for global markets created by technological change inevitably created a greater need for states to act together.

In both cases, the crucial agreement is constitutive as well as regulative.[19] In agriculture we can see this aspect most concretely in the soul of the Agreement, the Green Box,[20] whose general and specific criteria are designed to guide policymakers in the continuous process of policy reform. It contains those kinds of domestic support that will still be considered legitimate because they help farmers in one country without hurting farmers in another.

In telecommunications, it is the RP that critics see as regulative. Both the RP and the Green Box attempt to solve the same problem: when the economy is embedded in society, the social response to economic change will have locally specific characteristics. In the compromise of embedded liberalism, states accept that the structure of policy differs by country. The WTO does not impose either central regulation or a common regulatory framework. The challenge is always to understand and reconcile differences.

Notes

1 For the Punta Declaration, *see* Croome (1995).
2 A slightly different emphasis is adopted in Finnemore and Sikkink (1998, pp. 887–917).
3 On the WTO as the concrete embodiment of the regime, *see* Wolfe (1999a). For a more complete discussion of regime theory, *see* Wolfe (1999c).
4 For a description and quantification of the war, *see* Wolfe (1998, p. 63ff).
5 For details, *see* the Appendix to Wolfe (1998).
6 Reliable statistics are not available for any country prior to the mid-1980s, and the statistics are still problematic, especially for non-OECD countries.
7 Technically, they adopted the Report of the Group on Basic Telecommunications (WTO, 1997) which records the agreement to annex 55 schedules of commitments to the Fourth Protocol to the General Agreement on Trade in Services (WTO, 1996), itself a mechanism for adding to the Schedules of the GATS. For details, *see* Fredebeul-Krein and Freytag (1997, pp. 477–91).
8 It was also criticized for not going beyond what was happening anyway. Both views are represented in Drake and Noam (1998).
9 The technological and policy issues are surveyed in Cable and Distler (1995).
10 For a complete description, *see* the WTO website.
11 Source: ITU Direction of Traffic data reported in the *Financial Times* (1997).
12 Ruggie has frequently argued that the fact that the hegemon was American did much to shape postwar order.
13 *See*, for example, ITU (1995). For an assessment of the outcome, *see* ITU (1998).
14 The distinction was introduced to US law for purposes of delineating the extent of FCC jurisdiction over new telecommunications services based on data-processing. It is a relic of American regulatory structure now embedded in the RP and GATS which has little technological basis and is not really reflected in regulatory frameworks elsewhere. *See* Bronckers and Larouche (1997, pp. 16–18).
15 The Reference Paper, adopted by the Negotiating Group on Basic Telecommunications on 24 April 1996, may be found on the WTO website. It is also reproduced as an appendix to Bronckers and Pierre Larouche (1997).
16 For more on the Reference Paper, *see* Wolfe (1999b).
17 For an account of the changes in American policy and their effects on other countries, *see* Schultz and Brawley (1996).
18 EU policy did change on that date, but implementation will be slow – for example, in setting up new regulators to replace PTTs. For details, *see* Himsl and Milton (1998).
19 On this distinction, *see* Onuf (1989, pp. 50–2).
20 Article 7 is a description of the principles under which states may claim exemption from the general commitment to reduce domestic support policies for farmers. The

Green Box contains policies that 'have no, or at most minimal, trade distortion effects or effects on production'. Governments can do whatever they think necessary to support the income and general standard of living of farmers, but such support must not be related in any way to price levels or production volumes.

References

Bronckers, Marco C. and Pierre Larouche (1997) 'Telecommunications Services and the World Trade Organization', *Journal of World Trade*, vol. 31 (3), (June), pp. 5–48.

Cable, Vincent and Catherine Distler (1995) *Global Superhighways: the Future of International Telecommunications Policy*, London: Royal Institute for International Affairs.

Cerny, Philip G. (1991) 'The Limits of Deregulation: Transnational Interpenetration and Policy Change', *European Journal of Political Research* 19, pp. 173–96.

Cowhey, Peter F. (1990) 'The International Telecommunications Regime: the Political Roots of Regimes for High Technology', *International Organization*, vol. 44 (2), (Spring), pp. 169–200.

Croome, John (1995) *Reshaping the World Trading System: a History of the Uruguay Round*, Geneva: World Trade Organization.

Drake, William J. and Kalypso Nicolaïdis (1992) 'Ideas, Interests, and Institutionalization: "Trade in Services" and the Uruguay Round', *International Organization*, vol. 46 (1), (Winter) pp. 37–101.

Drake, William J. and Eli M. Noam (1998) 'Assessing the WTO Agreement on Basic Telecommunications', in Gary Clyde Hufbauer and Erika Wada (eds), *Unfinished Business: Telecommunications after the Uruguay Round*, Washington: Institute for International Economics, pp. 27–61.

Financial Times (1995) 20 September, p. 5.

Financial Times (1995) 4 October, p. 9.

Financial Times (1997) 'Review of the Telecommunications Industry', September.

Finnemore, Martha (1996) *National Interests in International Society*, Ithaca: Cornell University Press.

Finnemore, Martha and Kathryn Sikkink (1998) 'International Norm Dynamics and Political Change', *International Organization*, vol. 52 (4), (Autumn), pp. 887–917.

Fredebeul-Krein, Markus and Andreas Freytag (1997) 'Telecommunications and WTO Discipline: an Assessment of the WTO Agreement on Telecommunication Services', *Telecommunications Policy*, vol. 21 (6), pp. 477–91.

Himsl, Mike and Leslie Milton (1998) 'Making Good on Access Commitments: Implementation of WTO Commitments on Basic Telecommunications Services by the EU Member States, the United States, and Canada', paper delivered to the LSUC/CBA conference on New Developments in Communications Law and Policy, Ottawa, April.

ITU (1995) 'Trade Agreements on Telecommunications: Regulatory Implications', International Telecommunications Union: Report of the Fifth Regulatory Colloquium on the Changing Role of Government in an Era of Telecom Deregulation, December. Available on the web at http://www.itu.int/pforum/trade-e.htm.

ITU (1998) 'Trade in Telecommunications', International Telecommunications Union: Third Draft of the Secretary-General's Report to the Second World Telecommunication Policy Forum, 15 February. Available on the web at http://www.itu.int/wtpf/sg_rep/3_draft/3rd.htm.

Kratochwil, Friedrich and John Gerard Ruggie (1986) 'International Organization: a State of the Art on the Art of the State', *International Organization*, vol. 40 (4), (Autumn) pp. 753–77.

Langdale, John V. 'International Telecommunications and Trade in Services: Policy Perspectives', *Telecommunications Policy*, vol. 13 (3), (September 1989), pp. 203–23.

Murphy, Craig N. (1994) *International Organization and Industrial Change: Global Governance since 1850*, Cambridge: Polity Press.

OECD (1987) *National Policies and Agricultural Trade*, Paris: Organization for Economic Co-operation and Development.

OECD (1990) *Trade in Information, Computer and Communication Services*, Paris: Organization for Economic Co-operation and Development.

Onuf, Nicholas Greenwood (1989) *World of our Making: Rules and Rule in Social Theory and International Relations*, Columbia, SC: University of South Carolina Press.

Polanyi, Karl (1944) *The Great Transformation: the Political and Economic Origins of Our Time*, Boston: Beacon Press.

Ruggie, John Gerard (1983) 'International Regimes, Transactions, and Change: Embedded Liberalism in the Postwar Economic Order', in Stephen D. Krasner (ed.) *International Regimes*, Ithaca and London: Cornell University Press, pp. 195–231.

Schultz, Richard J. and Mark R. Brawley (1996) 'Telecommunications Policy', in G. Bruce Doern, Leslie A. Pal and Brian W. Tomlin (eds), *Border Crossings: the Internationalization of Canadian Public Policy*, Toronto: Oxford University Press, pp. 82–108.

Tarjanne, Pekka. (1997) 'Telecommunications and Trade', paper delivered by the Secretary-General of the International Telecommunications Union in Moscow, 5 February.

Vogel, Steven Kent (1996) *Freer Markets, More Rules: Regulatory Reform in Advanced Industrial Countries*, Ithaca, NY: Cornell University Press.

Wolfe, Robert (1998) *Farm Wars: the Political Economy of Agriculture and the International Trade Regime*, Basingstoke, Macmillan Press Ltd; New York, St. Martin's Press Inc.

—— (1999a) 'The World Trade Organization', in Brian Hocking and Steven McGuire, (eds), *Trade Politics: Environments: Issues: Actors and Process*, London and New York: Routledge.

—— (1999b) 'Regulatory Diplomacy: Why rhythm beats harmony in the trade regime', in Thomas J. Courchene (ed.) *Room to Manoeuvre? Globalization and Policy Convergence*, Kingston: John Deutsch Institute for the Study of Economic Policy.

—— (1999c) 'Rendering unto Caesar: How legal pluralism and regime theory help in understanding "multiple centres of power"', paper prepared for delivery to the Project on Trends Workshop on Multiple Centres of Power, Victoria, 13 May.

Woodrow, R. Brian (1991) 'Tilting towards a Trade Regime: the ITU and the Uruguay Round Services Negotiations', *Telecommunications Policy*, vol. 15 (4), (August), pp. 323–43.

WTO (1996) *Annual Report: Volume 1*, Geneva: World Trade Organization.

WTO (1997) *Report of the Group on Basic Telecommunications*, WTO: S/GBT/4, 15 February.

WTO (1996) *Fourth Protocol to the General Agreement on Trade in Services*, WTO: S/L/20, 30 April.

Zacher, Mark W. with Brent A. Sutton (1996) *Governing Global Networks: International Regimes for Transportation and Communications*, Cambridge: Cambridge University Press.

3
Institutional Learning in International Financial Regimes

Tony Porter

Often international finance is portrayed as a paradigmatic case of atomistic market interactions operating relatively free of institutional constraints. The apparent ability to move millions electronically at the touch of a button, the free-wheeling individualistic culture of global securities markets, and the desire and capacity of participants to escape state regulations all reinforce this image. This portrayal is, however, highly misleading. As this chapter will indicate, international finance involves and is dependent upon a dense fabric of institutions, ranging from informal social institutions, such as the private tacit understandings which have regulated the Eurobond markets, to large formalized organizations such as the International Monetary Fund.

Once the importance of the institutional dimension of international finance is recognized then one is led directly to ask how well these institutions work. It is readily apparent that international finance is hardly free from serious problems. Indeed the history of international finance is one of recurring crisis (Kindleberger, 1989; Porter, 1995; Suter, 1992), in the distant past, as with the Florentine crisis of 1340 triggered by the repudiation by the English king of international debts, through more recent crises, such as the global stock market crash of 1987, the European Monetary System crisis of 1992, the Mexican peso crisis of 1994, and the East Asian currency crises of 1997 and 1998. A number of features of international finance contribute to its susceptibility to crisis including: the degree to which intangible high-value transactions are dependent on fragile confidence and trust; the interdependence of financial actors which can lead to contagion in crises; and the centrality for and interpenetration of other sectors by the financial sector. Global finance is also marked by breathtaking change. Perhaps in no other issue area, then, is the question of learning and its relationship to institutional performance more pressing.

This chapter analyzes the capacity of international financial regimes to learn in two steps. First, I discuss the relevance of the notion of learning to international regimes in general and with respect to global finance overall. I then analyze the capacity of existing specific regimes for regulating international finance.

Social learning

The term 'social learning' has been used in several ways. Psychologists and social workers have developed *social learning theory* to analyze the role of imitation and social reinforcement in shaping behavior (Thyer and Wodarski, 1990). Hall (1993, p. 278) has used it to refer to learning that is centered in the state, and notes that 'we can define social learning as a deliberate attempt to adjust the goals or techniques of policy in response to past experience and new information'. Industrial economists have used it to analyze the way in which firms learn to use new technologies by 'observing the adoption process of other firms,' (Kapur, 1995, p. 173). Common to these uses is the idea that learning is not simply an individual process but is rather collective to an important degree, whether the collectivity is a society, an organization, such as the state, a set of organizations, such as the firms in an industry, or a more informal group of individuals.

It has also been pointed out that different levels of learning can be distinguished. The general tendency is to treat incremental adjustment of behavior in response to change as 'adaption' rather than learning (Nicolini and Meznar, 1995, p. 736; Haas, 1990).[1] Learning itself can range from low-level learning within a single context to high-level learning about causation, 'learning how to learn' (Nicolini and Meznar, 1995, p. 737, referring to Bateson, 1972), that allows the learner to learn and apply lessons across contexts. The editors of the present volume have similarly highlighted the importance of 'learning by learning'. Hall (1993) has distinguished between: the levels or settings of policy instruments (for example the amount spent); the techniques used to achieve policy goals; and a profound alteration in the 'policy paradigms', the taken-for-granted assumptions and world view within which policy is developed and implemented. Learning in the first two cases, he suggests, is an elite affair that occurs primarily within the state, while in the third case it involves widespread political conflict and discussion that spreads into society as a whole – indeed existing bureaucrats may well resist such learning.

Two conditions would appear to be important for successful and higher-level social learning. First, the capacity for the collection, storage, and dissemination of information must be well developed. This reflects the importance of communication and memory in learning and requires more than a minimal degree of institutionalization in order to facilitate the management of knowledge. Second, institutions are needed that facilitate reflective open deliberation and that encourage challenges to accepted orthodoxy rather than a defence of entrenched practices. It can be noted that these two conditions are not necessarily mutually reinforcing – indeed an increase in institutionalization may reduce the openness of deliberation. Both of these conditions can be problematic at the international level where institutions have traditionally been weaker as compared to institutions in particular nation-states.

In this chapter, then, I will assess the degree to which the international institutions that regulate global finance satisfy these conditions. I will make two arguments. First, international finance is sufficiently institutionalized to speak meaningfully about learning. Second, the system's capacity for higher level social learning has increased over time, but institutional fragmentation and lack of open deliberation remain serious problems. There are striking instances of system-wide learning, but these are sporadic and appear in the wake of crises.

Before examining particular institutions, it is important to acknowledge two complications in such analysis. First, it is difficult to clearly separate social learning in international finance from social learning in the international system more generally. Modelski (1990) has convincingly argued that there is 'evolutionary learning' in world politics. The lessons thereby learned inevitably involve international finance. In part this is because of the centrality and interconnectedness of finance to the economy as a whole, and in part it is because of the lack of any well-bounded financial organization that is responsible for the issue area as a whole, comparable, for instance, to a finance ministry domestically, which can develop relatively autonomously. Second, even leaving aside global learning, a comprehensive analysis of social learning in international finance would have to look at: learning within institutions, such as the IMF or the G-10, that aspire to develop policy for global finance as a whole; learning among states about how to regulate finance; learning among financial firms; and learning among individuals involved directly or indirectly in the international financial system, which could include market professionals, investors, and citizens affected by changes in global finance. Even where the focus is international regimes, as is the case for the present chapter, these other levels are relevant because regimes are interesting not just for their effects on those most directly involved in their management, but also for their potential effect on activity in the issue area as a whole. Moreover, as Hall points out, more profound types of learning may require interaction and conflict that extend beyond the confines of normal policy-making venues. Nevertheless it is necessary, given space constraints, to focus on systemic learning, which refers here to the capacity of the system as a whole to reflexively address problems, and to discuss only peripherally learning at these other levels.

Assessing learning in international financial regimes

This century has seen dramatic changes in the organization of international finance but also repetitions of past patterns. Most notable is a large-scale three-phase shift from highly internationalized securities markets organized around Britain in the late nineteenth century, to nationally based systems of regulation organized in conjunction primarily with bank-intermediated lending in the middle twentieth century, and finally a return to highly internationalized securities markets in the late twentieth century, this time organized mainly around the US (Porter, 1995). Accompanying this cyclical pattern is a secular trend towards greater institutionalization of global finance. I will briefly review this history, attempting to differentiate changes driven by power politics or a simple expansion of market forces from more reflexive initiatives which could be called social learning. In this first section, I will discuss learning in the international financial system generally. In the subsequent section, which focuses on the period since 1975, the institutional character of global finance had become much more differentiated and I shall therefore shift to an assessment of social learning in particular issue areas within global finance.

From British hegemony to 1975

In part, twentieth-century changes in the organization of international finance reflect the shift in the world system from British to US hegemony. This is evident not only in the shift in the center of global finance from London to New York, but also in the rise to pre-eminence of US-style forms of financial organization.

Under British hegemony international financial flows primarily took the form of foreign bonds, sold in London to a small number of wealthy investors, organized into the Committee of Foreign Bondholders, and mostly destined for large infrastructural investments carried out or regulated by the state in newly industrializing countries such as the US, Australia, Canada and Russia (Lipson, 1985). The gold standard provided an additional source of stability. The 1880s saw a crisis in this system (Suter, 1992). The symptom of the crisis was numerous defaults, but the deeper problem was the inadequacy of such an arm's-length system for monitoring the financing of increasingly technologically complex investments in increasingly peripheral locations.

By contrast US-based cross-border flows, which began to be significant in the first quarter of the twentieth century, were increasingly transferred through more hierarchical arrangements, whether this was through direct investment of non-financial multinational corporations or in the form of bank-intermediated foreign loans. Broad-based decentralized mobilization of funds in US stock markets to be used by non-financial multinational corporations was also an innovation that contributed to the displacement of the British-based system. The organizational advantages of the US-based system could justifiably lead one to discern evidence of learning, however such a view must be qualified: as Roe (1994) has argued, US-style corporate finance was in large part a political effect, expressed in New Deal and anti-trust regulation, of populist anger against the concentrations of economic power that were associated with the leading role in the early part of this century played by securities houses such as the Bank of Morgan. The international spread of this US-style of finance was also fostered by the deployment of the power of the US state after the Second World War that moved the corporate systems of its occupied major competitors, Germany and Japan, closer to its own.

The relatively insulated national financial systems that characterized the post-Second World War order can similarly be regarded as a mixture of learning and politics. After the First World War states as diverse as Russia and America began to be more sensitive to politically mobilized populations and to seek to insulate them from the painful adjustments associated with internationalized finance. This tendency accelerated with the great Depression of the 1930s and was a major motivation in the construction of postwar finance initiated in Bretton Woods in 1944. Helleiner (1994) has persuasively demonstrated that negotiators, including John Maynard Keynes for the UK and Harry Dexter White for the US, decided quite consciously to restrain cross-border capital flows which were seen as potentially destabilizing and harmful to the real economy. In general, the architects of the Bretton Woods system sought to chart a middle course between, on the one hand, the relatively autarkic nationally based policies of the 1930s with their destructive effects on the international economy and on the other hand, the excessive rigidity of the gold standard, which forced domestic populations to adjust to international rules. The resulting compromise has been termed, by Ruggie (1983), 'embedded liberalism'. Capital mobility was relatively restricted, exchange rates were pegged to the US dollar which was in turn pegged to gold, but were adjustable with the consent of the International Monetary Fund, and balance of payments financing was available from the IMF to cushion countries suffering from short-term fluctuations in international markets. A system of state-led development financing organized bilaterally and multilaterally, through the World Bank and other agencies, was put into place.

This new system certainly involved an increase in the capacity of systemic learning, evident in the Bretton Woods institutions, most notably the IMF, which involved an unprecedented organizational complexity in the governance of international finance and money, and in the open and innovative deliberative process which characterized their creation. Learning was stimulated by disagreements among its architects but also, and relatedly, by the political struggles of the interwar years which had contributed to policymakers' sensitivity to domestic opinion.

The breakdown of the post-Second World War arrangements, with their relatively insulated domestically based financial systems, in the face of globalization pressures is often attributed to the ability of market participants to outwit state regulators by devising new unregulated financial instruments and by playing one jurisdiction off against another to induce downward pressure on regulation. Others would supplement this account by emphasizing the growth in the institutional capacity of international markets – for instance, once recognized practices and firms were established for selling new international financial instruments, such as Eurobonds, then each subsequent sale becomes greatly facilitated. However the state is important too – Helleiner (1994) and others have pointed out that market-based explanations alone are unconvincing. In fact, states played a key role – either by taking particular actions or by deciding not to act to restrain markets – in bringing about the globalization of finance. In part, such state initiatives were driven by self-interest, as in the case of Britain's desire to reinvigorate London as a financial center after the Second World War by making it the centre of the Euromarkets.

Although there was a tremendous amount of market innovation involved in the re-emergence of globalized finance in the 1960s and early 1970s, it would be difficult to characterize this period as involving high degrees of social learning, particularly in comparison to the previous historical turning points noted above. Much of the innovation was based on the attempts of individual firms to maximize their own market position. Although the aggregate effect of these efforts was to establish new social practices and institutions, it was more unintended than reflexive. Despite the impassioned commitment of some economists to the liberalization of financial markets, there was much less deliberate collective reflection on the inadequacies of past practices and creative consideration of how jointly to devise solutions than was the case with the creation of the Bretton Woods system or of nationally based financial systems in response to the difficulties of the 1930s. Indeed the collapse of the Bretton Woods system was accompanied by a series of ad hoc and generally inadequate initiatives on the part of policy-makers, including: the 1968 creation of the Special Drawing Rights in an attempt to substitute for dollars, as trust in the latter currency dwindled; President Nixon's 1971 surprise unilateral abandonment of the US commitment to exchange dollars for gold, fatal for the Bretton Woods regime; and the hesitant period of recognition, from 1971 to 1976, that the system was dead.

By the mid-1970s a considerable wariness among states and markets had developed regarding the potentially negative consequences of rapid financial globalization. Initially these concerns centered on the potential damage to the real domestic economy and to international trade of highly volatile exchange rates and on the possibility that banks could fail as a result of their heightened involvement in high and poorly understood risks associated with international financial transactions. Moreover, the increasing international interconnectedness of banking systems and the downward pressure on prudential regulation increased the chances that financial distress could spread from one bank to another. Such concerns stimulated much reflection and institutional innovation and can therefore be seen as a stimulus to social learning. Additionally, the years of sustained international market and regulatory innovation had resulted in a sort of critical mass of joint financial practices and institutions that could be seen as providing a basis for increased reflexivity and social learning at a global level.

Social institutions and social learning in contemporary international finance

The above section has discussed the evolution of international finance and periods of social learning in very general terms. While such a broad-brush approach inevitably overlooks significant instances of social learning in particular issue areas, it does capture the most important features of the period before 1975. More recently, however, global finance has become much more institutionalized and differentiated. I therefore start with a discussion of system-steering initiatives and then look more specifically at functional issue-areas within global finance.

System-steering institutions

The most notable change at the system-steering level after the collapse of the Bretton Woods system was a shift towards more informal cooperative initiatives organized by the wealthiest countries through the G-7 and the G-10. This displaced to a degree both more formal and universal organizations such as the IMF and the UN, and unilateral initiatives, particularly on the part of the US.

The G-7 process was initiated by a 1975 informal meeting of the leaders of the US, the UK, France, Germany, Japan and Italy (Canada joined the following year) to discuss the problems associated with the collapse of the Bretton Woods monetary system and other economic concerns. Over the years, despite efforts to keep it informal, the process became quite institutionalized, with virtually year-round meetings and preparations by lower-level officials supplementing the annual summit meetings. Despite skepticism occasioned by the use of the meetings by leaders for photo-ops and rhetoric, the G-7 has increased its steering functions over the years. For instance Kirton (1993, p. 353) calculates that the number of 'specific instructions issued to component and outside bodies' increased from 6 at the first 1975 summit to 88 at the 1991 summit. Although the agenda was expanded to include security issues, the G-7 continues to play an important role in initiating alterations in economic policy.

Starting with the 1995 Halifax summit, which was held in the immediate aftermath of the Mexican peso crisis, the G-7 began to devote particular attention to the governance of the global financial system.[2] At that summit it was noted that there was a need to manage more effectively financial developments with potentially broad economic implications. . . . The international community must also improve its ability to address the risks inherent in the dramatic growth in private financial flows, the increased integration of domestic capital markets, and the greater recourse to financial innovations (G7, 1995, Section 5).

The summit called for an improved early warning system for financial shocks to be accomplished through stronger surveillance of national economies. It also called for heightened cooperation among regulatory authorities and a strengthening of regulation, in which the International Monetary Fund would play a more active role. Finally, it called for a streamlined and speedier IMF mechanism for disbursing funds in emergencies, as well as an increase in the funds available to be dispensed. The 1996 Lyons summit initiated a study of the use of electronic money for retail payments and called for a strategy for fostering financial stability in emerging market economies. Both the 1996 Lyons and 1997 Denver summits reviewed progress on the recommendations of the previous summits and reiterated their importance. The meetings of G-7 finance ministers, which have been integrated into the G-7 process since 1986, have played a particularly prominent role in the G-7's post-Halifax consideration of financial governance.

The G-10 central bank governors have also played a system-steering role in initiating committees and working parties to address various policy problems as they emerge.[3] As a Bank for International Settlements (1997) publication put it:

> The G-10 meetings have, over time, become the pivotal forum in which much wider activities have been set in motion by the G-10 central banks in pursuit of international financial stability.

Committees have included the Eurocurrency Standing Committee, established in 1971, which was given a mandate in 1980 to monitor international banking and produced an important report on international financial innovation in 1986, the Committee on Interbank Netting Schemes (established 1989), and others. The most prominent of these is the Basle Committee on Banking Supervision which is discussed further below. The Bank for International Settlements provides the Secretariat and meeting venue for the G-10 and its committees. The G-10 has also directly responded to the G-7's financial initiatives, most notably in the formation of working parties on financial stability in emerging economies and on retail electronic payments.

The heightened importance of the G-7 and G-10 in the post-Bretton Woods era does not mean that the more formal and universal organizations like the IMF and World Bank have diminished absolutely in strength

– on the contrary, as is well known, they too have increased their system-steering capabilities through promotion of structural adjustment policies in developing countries and through their production of statistical information and policy advice on financial matters, and indeed the G-7 has sought to increase the IMF's strength in recent years. Considering that one might have expected the IMF to disappear along with the Bretton Woods system it was designed to facilitate, it is remarkable that the organization was able to carve out a leading role for itself in the resolution of the debt crisis.

Nevertheless, the role of the IMF in the debt crisis can be seen as a reactive response to an existing problem that only secondarily affected the industrialized countries. By contrast, more proactive initiatives that primarily concern the industrialized countries are dealt with through the G-7 and G-10. This was even the case for financial instability in the G-10 countries that was related to the debt crisis, which was primarily handled at the G-10's Basle Committee on Banking Supervision. One can therefore say that the G-7 and G-10 are more central for system-steering than are the more formal and universal organizations such as the IMF and World Bank. The greater flexibility of the less formal G-7 and G-10 processes, the ability of the wealthiest countries to develop policies with minimal input and interference from the rest of the world, the presence at meetings of policy-makers with the power and knowledge to really make things happen, and the need of a relatively weaker US to coordinate more with its closest allies have all contributed to the enhanced role of the G-7 and G-10 in the post-Bretton Woods era. We will return to the significance for social learning of the G-7 and G-10's system steering initiatives after examining more specific regulatory arrangements in global finance.

Institutional capacity and social learning in specific issue areas in global finance

We now turn to an assessment of five regulatory areas in global finance. The first two of these are concerned with the international prudential regulation by states of banking and securities firms respectively. Prudential regulation refers to regulations designed to prevent financial crises by ensuring honest, competent, and careful management of financial firms. The next three areas are concerned with the liberalization of financial flows: under the North American Free Trade Agreement; under the negotiations initiated in the Uruguay Round; and under the emerging regime for the regulation of investments flows. This chapter's two central themes concerning social learning will become more apparent in this examination: it is clear that these areas are sufficiently institutionalized to speak of social learning and that this institutionalization has increased over time, but that a lack of coherence between these areas and a lack of open deliberation remains an impediment to social learning.

The prudential regulation of international banking and securities markets

In the post-Bretton Woods period, institutionalization and learning has occurred to an impressive degree in the international regulation of banking. As the globalization of finance proceeded during the 1970s and 1980s there was much concern that banks, by playing one national regulator off against another and threatening to flee to more lightly regulated jurisdictions, could induce a downward spiral in the regulatory standards, a 'race to the bottom', as all regulators would be forced to compete for bank business by reducing regulation. In response to concern about bank crises involving inadequate control and monitoring of international transactions, the G-10 central bank governors set up, in 1974, the Basle Committee, consisting of bank regulators from the G-10 plus Luxembourg. The Basle Committee, by facilitating cooperation among national regulators and by developing its own regulatory capacity, has prevented downward pressure on bank regulation, and indeed has strengthened international regulatory standards. A capacity for social learning has been an important factor in the Basle Committee's success.

The Basle Committee has succeeded in offsetting downward pressures on regulation by establishing channels for closer cooperation among regulators, including a clearer division of labor, improved procedures for sharing information, and agreed and more explicit minimum standards. Such efforts were first set out in a Concordat issued in 1975 and substantially revised and supplemented in 1983 and 1992. The 1983 Concordat, for instance, established the principle of 'consolidated supervision' – an agreement that a home regulator should obtain information on, and supervise, all the operations of the banks headquartered in its jurisdiction. This was an effective tool in regaining the initiative relative to banks as it reduced the ability of banks to escape regulation by shifting activities to foreign offices. In 1992, a more explicit set of standards for adequate supervision was set out and regulators agreed to deny entry to banks that did not meet these standards.[4]

One particular set of standards which has been the centerpiece of the Basle Committee's work are capital adequacy standards, first issued in July 1988. Increasing bank capital, because capital can be regarded as the difference between assets and liabilities, provides a bigger cushion against bankruptcy, but also, because it represents shareholders' equity, imposes more risk and responsibility on shareholders, increasing their incentive to make sure managers do a good job. The Committee's capital standards are technically complex, having taken the Committee years to model and test before their initial establishment, and involving continual efforts to refine them since 1988.

International securities regulation remains much less effective than bank regulation. The counterpart to the Basle Committee is the International Organization of Securities Commissions (IOSCO) in Montreal, the membership of which includes 152 state and non-state securities regulators (IOSCO, 1997a, p. 14). Although IOSCO appears more impressive than the Basle

Committee on paper, in terms of its character as a formalized and differentiated bureaucratic organization, it has not displayed the cohesion and forward momentum of the Basle Committee. For instance, it has lagged behind the Basle Committee in the establishment of capital adequacy standards: indeed, the modest progress that was made on these standards was achieved after the Basle Committee began working with, and putting pressure on, the securities regulators at IOSCO. IOSCO continues, to a much greater degree than the Basle Committee, to rely on, or have overlapping responsibilities for regulation with, private institutions, such as the International Federation of Stock Exchanges (known by its French acronym FIBV) which sees itself as working with the regulators at IOSCO but at the same time as countering excessive state regulation by fostering private self-regulatory initiatives (FIBV Annual Report, 1994). Despite these weaknesses, IOSCO has produced many significant technical reports on the international regulation of securities markets. In addition to identifying problems, these play a coordinating role by fostering a common direction among securities regulators, including regulators in emerging markets.[5]

Along with the emergence of state-initiated arrangements for regulating international banking and securities markets has come a significant institutionalization of the private sector's capacity for self-regulation. For securities markets, which have always tended to rely more heavily on private self-regulatory arrangements than is the case for international banking, the leading private institution is the FIBV. The shared commitments that the FIBV represents are remarkably formalized, including: 22 statutes; 19 internal rules organized into 5 articles; 'General Accepted Principles with Respect to Securities Trading', which were adopted in 1988 with which all members must comply; 'Generally Accepted Principles of Securities Business Conduct'; and six organized committees and subcommittees.

The Group of Thirty, established in 1978, bills itself as a 'private, independent, nonpartisan, nonprofit body' (G-30 Annual Report, 1994, 3). Its 1994 30-member composition is typical: chaired by Paul Volcker, former Federal Reserve Chair, it includes top private bankers (for example Deutsche Bank, Goldman Sachs, Yamaichi Securities), central bankers and finance officials (Bank of England, Banque de France, Banca d'Italia), and academics (Peter Kenen, Paul Krugman, Sylvia Ostry). It seeks to influence the development of financial markets by issuing reports and running seminars and played a more practical role in spearheading efforts to improve clearance and settlement practices in the world's securities markets.

For commercial banks the most prominent international association is the Institute of International Finance. The roots of this association lie in the debt crisis: the organizing committee that founded it was established by representatives of 30 banks in October 1982. By 1990, it had a staff of 40, a $5 million budget, 147 full member banks and financial organizations from 39 countries, representing over 80 per cent of international bank exposure in the developing world (Sarver, 1990, pp. 433–7). Its primary function has been to share information about sovereign borrowers but it has more recently begun developing positions on regulatory matters.

The liberalization of financial flows

In addition to institutions concerned with prudential regulation there has been a dramatic strengthening of the international institutions governing the liberalization of financial flows, contrary to the view that freer trade means a reduced role for the state. In this section, we look at three key examples of this strengthening: the NAFTA, the Uruguay Round and the emerging regime for regulating investment, each of which have created a whole new set of trade rules and procedures.

The NAFTA

The NAFTA is especially important because it represents the most ambitious attempt to integrate developed and emerging markets. Moreover US negotiators hoped that the NAFTA's unprecedented strengthening of investment provisions would become the model for other financial services negotiations, including those in the Uruguay Round.[6]

The NAFTA's key chapters that are concerned with financial integration are 14 (Financial Services), 11 (Investment), and 20 (Dispute Resolution), along with the various Annexes that list reservations and exclusions. A sign of the strength of the agreement is that everything not in these lists is to be covered under the NAFTA.[7] The following provisions are particularly noteworthy.[8] First, financial firms are guaranteed the better of national treatment and most-favored nation treatment. Second, firms can carry out cross-border sales of financial services but not necessarily promote those sales in the foreign market. Third, rights regarding foreign ownership of financial firms are expanded and ensured over a number of years, a process which was accelerated on the Mexican side in 1995 in response to the negative effect of the peso crisis on Mexican banks. Fourth, requirements on foreign investments to source locally, have nationals on boards of directors, meet export targets, or restrict repatriation of profits and capital were renounced. Mexico renounced its long-standing position on national sovereignty with respect to expropriation, a position for which it had been the leading advocate in previous North–South negotiations, by agreeing that expropriation must be compensated promptly at 'fair market value' in convertible currency, precluding the use of arguments about rectifying past exploitative practices (Gantz, 1993).

One of the most remarkable features of the NAFTA is the provision, in Chapter 11, of the right for private investors to use binding arbitration to collect damages from or reverse decisions of states that do not comply. For the first time, Canada and Mexico committed themselves to use the International Center for the Settlement of Investment Disputes at the World Bank, the dispute resolution process favored by the United States.[9]

NAFTA's significance goes beyond its three current members because of the stated goal of bringing in other countries in the hemisphere and because Mexico's key commitments on financial services were extended to the North American subsidiaries of non-NAFTA firms.

The GATS

The commitments made to liberalization under the financial services negotiations in the Uruguay Round and subsequently in the WTO are weaker than in the NAFTA but this is offset by the greater number of participants in the WTO.[10] As with the NAFTA the WTO commitments revolve around national treatment and MFN principles. In both agreements, financial services were integrated into the trade negotiation process by first defining them and specifying the ways in which they could be delivered. As with services in general this is significant because prior to the Uruguay Round there was no consensus that services could be treated as tradeable goods and therefore be subject to trade liberalization principles (*see* the chapter by Robert Wolfe in this volume). As with the NAFTA the right of service providers to establish offices in host countries was expanded along with the strengthening of the rights of foreign investors in general. For instance, the strengthening of prohibitions on such trade-related investment measures (TRIMs) as trade balancing, minimum export requirements, investment incentives linked to exports, and local content requirements expands to some degree the rights and flexibility of foreign investors.

The regime for regulating investment

The outlines of an emergent multilateral regime for regulating investment flows are evident in the series of bilateral and regional investment treaties that now total more than 600 (Ostry and Winham, 1995: 77) and most of which involve one industrialized and one developing country. These bilateral arrangements were initiated by the Europeans and it was not until the late 1970s that the US began developing its own program (Shenkin, 1994: 573–82). The European agreements generally include reciprocal national and MFN treatment for existing investments and guarantee at least full compensation for nationalization. As of 1992 there were more than 400 of these European agreements. The US bilateral investment treaty (BIT) program seeks to go beyond the European one by covering the establishment of new investments and by covering all investment measures. As of 1992, however, only 25 of these US BITs had been signed (Shenkin, 1994, p. 579). Despite this small number they are significant because they are seen by the US government as expanding the basis in international law for an investment treaty along the lines preferred by the US (DeLuca, 1994). Both the Association of Southeast Asian Nations (ASEAN) and the Organization of the Islamic Conference have signed similar regional investment treaties (Shenkin, 1995). All these investment treaties generally provide for arbitration between investors and states at the International Center for the Settlement of Investment Disputes at the World Bank, creating considerably expanded rights for private investors relative to states.

In 1995 the industrialized countries, led by the European Union, began to campaign more vigorously to bring investment into the WTO. This would dramatically strengthen the more ad hoc regime that presently exists by linking investment liberalization with WTO membership and trade issues more generally: pressure to comply would increase sharply for those countries not presently supportive. Alarm was expressed at the EU initiative by NGOs and developing countries, most notably India and Brazil, both because of the perceived negative impact that a strengthened regime would have on the capacity of southern states to foster development, and because of concerns at the degree to which the negotiation process is dominated by the North (Third World Network, 1995).

In part because of US and EU concern at the resistance that investment negotiations would meet in the WTO, since 1995 the primary forum for creating a new Multilateral Agreement on Investment (MAI) has instead been the Organization for Economic Cooperation and Development (OECD), comprising the world's 28 most industrialized states. If negotiations are completed, non-OECD countries would also be able to sign the MAI. The agreement would govern investments in both tangible and non-tangible assets, would include most-favored nation (MFN) and national treatment provisions, would have strong provisions regarding compensation for expropriation and the free repatriation of profits, would permit the unrestricted transfer of key personnel by corporations across borders, and would provide for binding arbitration between states and investors that can be initiated by investors (Witherell, 1997; Council of Canadians, 1997).

Opposition to and concern about the MAI increased through 1997 and the negotiations proved to be more difficult than anticipated. By March 1998, it became apparent that the negotiations were not going to be completed in 1998 as planned and by 1999 it was clear that the negotiations were dead. It was likely, however, that similar negotiations would be revived at the WTO.

Assessing social learning in contemporary international finance

As noted at the outset of this chapter, social learning requires sufficient organizational capacity for managing flows of information, but also requires institutionalized opportunities for open deliberation and contestation. In assessing the degree to which these conditions are satisfied in the regulation of global finance I am making two arguments: first, international finance is sufficiently institutionalized to speak meaningfully about learning; second, the system's capacity for higher level social learning has increased over time, but institutional fragmentation and lack of open deliberation remain a serious problem: instances of system-wide learning are sporadic and appear in the wake of crises.

The above sections have indicated that there is a considerable degree of institutional development in international finance, contrary to those who portray it as consisting simply of atomistic market actors. This ranges from institutions such as the G-7 which seek to govern the system as a whole, to the micro-level market practices that facilitate cross-border financial flows. In this century as a whole, and since the collapse of Bretton Woods in particular, we have seen a greatly increased capacity for communicating, producing, assessing and storing information. The establishment of the Bretton Woods system itself, and of the International Monetary Fund, was a dramatic step in the development of the governance of international finance that explicitly sought to correct the problems of the interwar period and of the gold standard of the late nineteenth century. The various international institutions that have emerged since the mid-1970s have further enhanced the potential for systemic learning by providing an ongoing capacity to respond to new challenges, by producing highly technical reports and by fostering consensus on regulatory initiatives based on those reports. This is particularly evident in the G-10 and the other institutions concerned with prudential regulation. Each of the initiatives of the Basle Committee for Banking Supervision, for instance, has built on assessments of previous initiatives. The 1975 Concordat establishing a general division of responsibility between home and host regulators was revised and updated in 1983 and then again in 1992, each time in response to the lack of specificity and effectiveness of previous guidelines. Similarly, the 1988 capital adequacy agreement, itself the product of years of intensive research and modeling on the part of the Committee, has been subject to constant attempts to refine it and to bring new risks into its purview.

Despite this growth in institutional capacity for social learning, there remains a serious problem with lack of integration and coherence. While the G-7 and the G-10 seek to coordinate the system as a whole, their lack of strong organizational capacity relative to the number of matters on their agendas prevents them from engaging in sustained consideration of the system as a whole and there continues to be a haphazard quality to their initiatives. The various areas which were discussed above have developed institutions and policies with relatively little capacity to address the impact of developments in one area on another. The Financial Stability Forum established in 1999, while a step forward in addressing this problem, was remarkably weak given the severity of the crisis that preceded its formation. Especially problematic is the lack of formal mechanisms for integrating emerging market regulators into its work.

This is most striking in the lack of coordination between the institutions concerned with prudential regulation and those concerned with the liberalization of financial flows (Porter, 1997). Although both the NAFTA and the Uruguay Round agreements on financial services provided a 'prudential carve-out' – a provision allowing prudential concerns to supersede trade commitments – this was not accompanied by any enhanced capacity for prudential regulation. Prior to both the Mexican peso crisis of 1994 and the East

Asian currency crisis of 1997, countries were opened up to large financial inflows with inadequate consideration of the need to strengthen prudential regulation. In both crises, investors' swing from excessive optimism to excessive pessimism would have been offset had strong rules enhancing the transparency of local markets been in place or if investors' incentives to monitor their investments had been increased by restrictions on their ability to exit at a moment's notice.[11] The inadequacy of prudential regulation also greatly increased the severity of the crisis as weak banks became vulnerable to collapse when credit became scarcer. The centrality of inadequate prudential regulation in these crises is now well recognized, and although the IMF now addresses both prudential regulation and liberalization of financial flows, lack of coordination between the institutions involved in financial liberalization and those involved in prudential regulation and crisis management remains a serious problem.

This lack of coordination has also been evident in the relationship between banking and securities regulation. The Basle Committee developed its key regulations independently of the securities regulators at IOSCO even though the distinction in practice between the banking and securities industry was becoming weaker, meaning that banks were increasingly engaged in securities market activities and loosely regulated securities markets could channel funds away from better regulated banks. Some important recent progress has been made in coordinating among different types of financial regulators, first through joint efforts by the Basle Committee and IOSCO to coordinate in the development of capital standards, then through the creation in 1996 of the Joint Forum by the Basle Committee, IOSCO and the International Association of Insurance Supervisors,[12] and finally with the creation of the Financial Stability Forum.

There is also a need for institutional mechanisms to foster more open deliberation and contestation in the governance of global finance. There are examples of excessive secrecy. For instance, the Basle Committee has provided virtually no information on its deliberations except for the formal research and reports on standards that it publishes. It refused to provide a significant number of documents requested by a General Accounting Office team mandated by the US Congress (Porter, 1993). The Multilateral Agreement on Investment was negotiated with very little publicity: NGOs and legislatures were only reluctantly involved in discussion of it at a late date (Council of Canadians, 1997). The effects of secrecy are reinforced by the way in which the highly technical nature of global financial regulation makes it difficult for lay observers to understand the significance of decisions.

Where deliberation extends beyond regulatory institutions, it is private institutions, such as the FIBV, which are closely linked to the industry itself, that become involved. With the exception of those concerned about the Third World debt crisis and the MAI, there has been no other significant NGO community that monitors and discusses developments in the governance of global finance.[13] This restricts the range of perspectives that is brought to bear on problems, impedes the expression of views by non-practitioners who may nevertheless be severely affected by changes in global finance, and makes it difficult for broader questions about the role of global finance in societies more generally, to be considered. In other highly technical issue areas, such as arms control or the environment, an intermediate literature, produced by non-governmental organizations, has emerged to allow lay actors to monitor and have input into important policy questions. This has not occurred in global finance.

The centralization of policy-making initiatives in the G-7 and G-10 reduces the capacity for systemic social learning with regard to the role of developing countries in global finance. The Basle Committee has encouraged the creation of an Offshore Group as well as regional groupings of supervisors from emerging markets just as IOSCO has sought to enhance interaction between its Emerging Markets Committee and its centrally important Technical Committee. However the function of these organizations of developing country supervisors is as much to implement new regulations devised in the corresponding process dominated by industrialized countries as to initiate responses to questions of particular concern to emerging markets. The inadequacy for developing countries of existing regulatory arrangements is glaringly revealed by the contribution of problems in the banking sectors to the Mexican and East Asian crises noted above.

It is also clear that social learning is particularly pronounced in the wake of crises, and as such is reactive rather than proactive. The creation of the Basle Committee at the end of 1974 was triggered by the international risks that contributed to and resulted from the collapse of the Herstatt and Franklin Banks earlier in that year. The debt crisis of the early 1980s and the Mexican peso crisis contributed to further waves of research and institutional innovation. The 1997–98 crisis led to the creation of the Financial Stability Forum.

A final deficiency concerns the over-integration of short-term market-oriented practices and institutions relative to the more complex institutions that are needed for systemic learning. International financial markets offer stunning examples of the capacity of market participants to learn: new financial products proliferate at exponential rates. Yet they also provide stunning examples of the shortsightedness of market participants when it comes to the aggregate effect of their activities. There is a long list of activities into which financial firms have rushed, only to suffer from overcapacity and crisis a few years later: currency speculation in the wake of the collapse of Bretton Woods; the Third World debt crisis; the real estate boom and bust of the 1980s; the stock market boom of the 1980s and the crash of 1987; the Mexican peso crisis of 1994 and the East Asian crisis of 1997 and 1998. In each case, the negative effects of these crises have extended well beyond the firms and investors directly involved in them.

The market can often devise partial solutions to each of these problems but the solutions require greater sustained institutional capacity than is possessed by individual firms linked in arm's-length transactions. For instance, much of the impetus for the growth of financial derivatives is to supply non-financial firms with insurance against volatility in exchange and interest rates: by buying a financial future or an option, a firm can pay a premium and lock in a particular rate or establish a limit to the losses that can be expected from future fluctuations in the prices of financial assets. Unfortunately a rapid growth of derivatives, the prices of which are driven by expectations of their future movement rather than analysis of economic fundamentals and may even wash out price signals from the real economy, can introduce more instability into the system. In part, this is a classic collective action problem: it is not in the interest of any particular actor to invest in costly research and analysis of larger and more fundamental trends if it cannot appropriate the returns from that research and thus the supply of such knowledge to the market as a whole will be suboptimal. Moreover, stability requires authority – the acceptance of information and the compliance with a prescribed direction based on trust. Although particular firms, such as debt-rating agencies, manage to produce this type of knowledge and authority to a degree, it is often too little and too late, as evident in the frequency of crisis in international finance.

In conclusion, then, there has been considerable social learning in international financial regimes. However there is much more that needs to be done before we can say that the institutions which organize and govern international finance are equal to their task.

Notes

1   Haas (1990) has done the most to analyze learning in international organizations. Haas defines learning as 'the process by which consensual knowledge is used to specify casual relationships in new ways so that the result affects the content of public policy' (1990, p. 23). Learning organizations deal proactively with change by incorporating new problems into integrated conceptual frameworks that build on previous work. Haas suggests that a capacity to synthesize or transcend two opposing perspectives on a problem is a key feature of learning. Adaptation, by contrast, can involve marginal adjustments or the cooptation of challenging actors. Adaptive organizations may simply drop problems off their agendas as new problems emerge with no alteration of their fundamental causal or normative beliefs.
2   This was not the first time the G-7 had considered the regulation of global finance. For instance, the 1989 Paris Summit established the Financial Action Task Force on Money Laundering (FATF). *See* FATF 1997. Information on the G-7 can be obtained from the University of Toronto's G7 Research Group website (utl1.library.utoronto. ca/disk1/www/documents/g7).
3   The G-10 traces its origins back to 1962 when ten wealthy IMF members plus Switzerland agreed to provide additional funds to the IMF beyond their quotas, the 'General Arrangements to Borrow'. The G-10 arrangement allowed them to exercise

more control over this lending than if it had simply been processed through standard IMF channels. Further information about the G-10 and its Committees is available in the BIS Annual Reports and at the BIS website (*www.bis.org*).

4 For information on the work of the Basle Committee *see* various issues of *Report on International Developments in Banking Supervision* (RIDBS) published by the Basle Committee, and Porter (1993).

5 A concrete example of social learning in IOSCO is its November 1997 'Report on the Self-Evaluation Conducted by IOSCO Members Pursuant to the 1994 IOSCO Resolution on *"Commitment to Basic IOSCO Principles of High Regulatory Standards and Mutual Cooperation Assistance"* ' (available at www.iosco.org). This report identifies problems in the capacity of members to implement the 1994 Resolution and makes recommendations for addressing those problems.

6 This section on the NAFTA draws heavily from Porter, 1997.

7 By contrast both the CUSFTA and the GATS specified which financial services were covered by the agreement – all others were not. A useful summary of the financial services provisions is Bachman, Benedict and Analdúa, 1994.

8 The provisions that are discussed in this paragraph are subject to some exceptions. A more detailed understanding of the financial services provisions can be obtained from the NAFTA text itself or from Bachman, Benedict and Anzaldúa (1994).

9 It should be noted that disputes over the right for prudential financial regulation to deviate from NAFTA commitments and the Chapter 14 financial services agreements are not covered by Chapter 11 but by Chapter 20 binding interstate arbitration procedures and by a new Financial Services Committee, although the vague specification of the latter under Chapter 14 casts doubt on its likely effectiveness (McNevin, 1994).

10 This discussion of the GATS and the next section on the regime for regulating investment are drawn from Porter, 1999.

11 The advisability of restricting cross-border capital flows remains a matter of dispute. Nevertheless there are indications, in the wake of these crises, of an increased recognition of the potential benefits among those previously hostile to them (for instance IMF, 1995 and *Economist*, 1998).

12 The Joint Forum has been attempting to improve the exchange of information and 'examining ways to enhance supervisory coordination' across these industries. Joint Forum Press Statement, 27 Nov. 1997, available at www.bis.org/press. Some coordination had occurred previously under the auspices of the Joint Forum's predecessor, the Tripartite Group.

13 This statement is based on ongoing but unpublished research I am conducting into global civil society and global finance and can be partially confirmed by a review of the financial listings in the *Yearbook of International Organizations*.

References

Bachman, Kenneth L, Scott N. Benedict and Ricardo A. Anzaldúa (1994) 'Financial Services under the North American Free Trade Agreement: an Overview', *International Lawyer*, vol. 28 (2), (Summer), pp. 291–312.

Bank for International Settlements (1997) 'The Bank for International Settlements: Profile of an International Organization', Section 5, 'The BIS as a Forum for International Monetary Cooperation', at www.bis.org/about.

Bateson, G. (1972) *Steps toward an Ecology of Mind*, S. Albany, Australia: Paladin.

Council of Canadians (1997) 'When Corporations Rule the World', *Action Link* (July), Update, p. 1.

DeLuca, Dallas (1994) 'Trade-Related Investment Measures: US Efforts to Shape a Pro-Business World Legal System', *Journal of International Affairs*, vol. 48 (1), (Summer), pp. 251–77.

Dobson, Wendy (1991) *Economic Policy Coordination: Requiem or Prologue?* Washington: Institute for International Economics.

Economist (1998) 'Kill or Cure?', distributed on the Internet.

FIBV Annual Report, 1994.

Filipovic, Miroslava (1994) 'A Global Private Regime for Capital Flows', unpublished paper, York.

Financial Action Task Force on Money Laundering (1997), web page at www.oecd.org/faft/index.htm.

Gantz, David A. (1993) 'Resolution of Investment Disputes under the North American Free Trade Agreement', *Arizona Journal of International and Comparative Law*, vol. 10 (2), pp. 335–48.

Group of Seven (1995) 'Halifax Summit Review of the International Financial Institutions' Background Document, 16 June, available at the G-7 website at the University of Toronto, www.library.utoronto.ca.

Haas, Ernst (1990) *When Knowledge is Power: Three Models of Change in International Organizations*, Berkeley: University of California Press.

Hall, Peter (1993) 'Policy Paradigms, Social Learning, and the State: the Case of Economic Policymaking in Britain', *Comparative Politics*, vol. 25 (3), (April), pp. 275–96.

Helleiner, Eric (1994) *States and the Reemergence of Global Finance: from Bretton Woods to the 1990s*, Ithaca and London: Cornell University Press.

International Monetary Fund (1995) *International Capital Markets: Developments, Prospects, and Policy Issues*, Washington: IMF, August.

International Organization of Securities Commissions (1997a) 'Final Communiqué of the XXIInd Annual Conference of the International Organization of Securities Commissions', available at www.iosco.org.

International Organization of Securities Commissions (1997b) 'Report on the Self Evaluation Conducted by IOSCO Members Pursuant to the 1994 IOSCO Resolution on "Commitment to Basic IOSCO Principles of High Regulatory Standards and Mutual Cooperation and Assistance"', November, available at www.iosco.org.

Kapur, Sandeep (1995) 'Technological Diffusion with Social Learning', *Journal of Industrial Economics*, vol. XLIII (2), (June), pp. 173–95.

Kindleberger, Charles P. (1989) *Manias, Panics and Crashes: a History of Financial Crises*, New York: Basic Books.

Kirton, John (1993) 'The Seven Power Summit as a New Security Institution', in David Dewitt, David Haglund and John Kirton (eds), *Building a New Global Order: Emerging Trends in International Security*, Toronto: Oxford University Press, pp. 335–57.

Lee, Ching-Mei and Lorraine G. Davis (1993) 'The Study of Social Learning and Social Bonding Variables as Predictors of Cigarette Smoking Behaviour among Ninth-Grade Male Students in Taipei, Taiwan, the Republic of China', *Research in Human Capital Development*, vol. 7, pp. 115–34.

Lipson, Charles (1985) *Standing Guard: Protecting Foreign Capital in the Nineteenth and Twentieth Centuries*, Berkeley: University of California Press.

McNevin, Valerie J. (1994) 'Policy Implications of the NAFTA for the Financial Services Industry', *Colorado Journal of International Environmental Law and Policy*, vol. 5 (2), (Summer), pp. 369–99.

Modelski, George (1990) 'Is World Politics Evolutionary Learning?', *International Organization*, vol. 44 (1), Winter, pp. 1–24.

Nicolini, Davide and Martin B. Meznar (1995) 'The Social Construction of Organizational Learning: Conceptual and Practical Issues in the Field', *Human Relations*, vol. 48 (7), (July), pp. 727–46.

Ostry, Sylvia and Gilbert R. Winham (1995) 'Post-Uruguay Round Trade Policy', in Sylvia Ostry and Gilbert R. Winham (eds), *The Halifax G-7 Summit: Issues on the Table*, Halifax: Centre for Foreign Policy Studies.

Porter, Tony (1993) *States, Markets and Regimes in Global Finance*, Basingstoke: Macmillan.

—— (1995) 'Innovation in Global Finance: the Impact on Hegemony and Growth since 1000 AD', *Review*, vol. XVIII (3), (Summer), pp. 387–429.

—— (1997) 'NAFTA, North American Financial Integration and Regulatory Cooperation in Banking and Securities', in Geoffrey R.D. Underhill (ed.), *The New World Order in International Finance*, Basingstoke: Macmillan, pp. 174–92.

—— (1999) 'The Transnational Agenda for Financial Regulation in Developing Countries', in Leslie Elliott Armijo (ed.), *Financial Globalization and Democracy in Emerging Markets*, Basingstoke: Macmillan, pp. 91–114.

Roe, Mark J. (1994) *Strong Managers, Weak Owners: the Political Roots of American Corporate Finance*, Princeton: Princeton University Press.

Ruggie, John Gerard (1983) 'International Regimes, Transactions and Change: Embedded Liberalism in the Postwar Economic Order', in Stephen Krasner (ed.), *International Regimes*, Ithaca and London: Cornell University Press, pp. 195–232.

Sarver, Eugene (1990) *Eurocurrency Market Handbook*, 2nd edition, New York: New York Institute of Finance.

Sassen, Saskia (1995) 'When the State Encounters a New Space Economy: the Case of Information Industries', *American University Journal of International Law and Policy* (Winter), vol. 10 (2), pp. 769–89.

Shenkin, Todd S. (1994) 'Trade-Related Investment Measures in Bilateral Investment Treaties and the GATT: Moving toward a Multilateral Investment Treaty', *University of Pittsburgh Law Review*, vol. 55 (2), (Winter), pp. 541–606.

Suter, Christian (1992) *Debt Cycles in the World Economy: Foreign Loans, Financial Crises and Debt Settlements, 1820–1990*, Boulder: Westview.

Third World Network (1995) package of materials sent over the Internet by Martin Khor, Director of the Third World Network (TWN) on 9 November 1995. *See*, for instance, 'WTO Should not Negotiate Foreign Investment Treaty – NGO Statement on the Foreign Investment Issue in WTO', a statement signed by 32 NGOs. The TWN can be reached on the Internet at twn@igc.apc.org or twnpen@twn.po.my.

Thrift, Nigel (1994) 'A Phantom State? The De-traditionalization of Money, the International Financial System and International Financial Centres', *Political Geography*, vol. 13 (4), (July), pp. 299–327.

Thyer, Bruce A. and John S. Wodarski (1990) 'Social Learning Theory: toward a Comprehensive Conceptual Framework for Social Work Education', *Social Science Review*, vol. 64 (1), (March), pp. 144–50.

Underhill, Geoffrey R.D. (ed.) (1997) *The New World Order in International Finance* Basingstoke: Macmillan, 'Conclusion', p. 315.

Witherell, William H. (1997) 'Developing International Rules for Foreign Investment: OECD's Multilateral Agreement on Investment', *Business Economics*, vol. XXXII (1), (1 January), pp. 38–43.

4

Institutional Learning in Standards Setting

Liora Salter

Introduction

Much of the literature cited in this volume is preoccupied with changes brought about by the new economy. Its authors want to know whether and how institutions (and the rules, practices and conventions and social relations upon which they rest) are, or could be made congenial with this new economy and vice versa. The tenor of arguments is often prescriptive, and the focus of prescriptions is sometimes on government policy. Put prescriptively, the question becomes whether new institutions can be established, and old ones reformed, to support the kind of collaborative relationships necessary to meet the challenges of the new economy. A criterion for success is whether such new or reformed institutions learn, not only from their environment – which is rapidly changing – but also from themselves by self-reflection. It is generally believed that institutions will have to be self-reflective if the challenges of the new economy are to be met, but so, too, the new economy must develop in particular ways if it is to sustain both growth and democratic social relations.

There is much new and exciting about this conception of economic development and institutions. It emphasizes the social aspects of economic development, substituting a sociology of institutions for the psychology of individual motivation usually preferred by economists. Institutions are understood to be worthy of study in their own right, not just as adjuncts to individual behavior. Conventions and informal law are properly understood to penetrate to the core of formal rules and organizational practices, influencing their meaning and import in profound ways. Moreover, none of this is seen to be static, nor are institutions understood to be without internal contradictions. Institutions are capable not just of learning, but of doing and saying quite contrary things simultaneously. They are, at once, forces for learning and centres of rigidity.

Few here or anywhere (outside the popular press) would seriously claim that the new economy is entirely new. What has changed is the qualitative import of an increasing number of quantitative changes. At some point, an invisible (and somewhat arbitrary and fungible) line has been crossed between an old order and a new one. Many instances of the old remain intact, however, and the new developments are not so pervasive yet that they totally rewrite the basic logic of advanced industrial economies. Simply, economies are on the cusp of change, pulled in both directions at once. Change is likely to be more far-reaching than even prophets of cyberspace envision, but its course will not be uniform, consistent, necessarily progressive or unidirectional. The goal of this volume, then, is to understand some changes wrought by the new economy in their own terms but also as they meet, or fail to meet, the challenge identified above.

What is the challenge of the new economy actually? Not everyone is enamoured of the new economy on its own terms. Whatever economic benefits the new economy might deliver, almost everyone agrees that it threatens to exacerbate the cleavages between rich and poor, developed and underdeveloped. Further, the new economy potentially increases the quantum of alienation characteristic of advanced capitalistic societies by restructuring workplace relations, dislocating workers, creating a new and ever larger underclass, and stripping the state of its formerly flawed but nonetheless essential roles as a democratic forum, and in providing a safety net for all its citizens. Seen in this light, institutions must actually meet two challenges to be worthy of the label 'learning institutions'. They must demonstrate creative adaptation within the new economy on its terms, and creative (i.e., democratic and humane) response to the pressures introduced by the new economy and its negative by-products.

To explore how change is actually unfolding in the new economy, it is obviously useful to examine domains of economic activity clearly on the cusp of change. The domain of interest in this chapter is standardization for the new communication and information technologies, almost by definition, on the cusp of change. Needless to say, without standards – without agreed-upon technical capacities for the new technologies and their interconnection – there is no new economy. Indeed, the notion of a new economy presupposes an eventually fully interconnected world where new linkages become crucial because of the speed at which information travels, the rapid development and deployment of new technologies, new work practices and attendant relations, and because of the relative autonomy of economic development and locality. This fully interconnected world is premised upon the existence of standards for interconnection and for assuring harmonization throughout. When the new management practices standards are added to technical standards, standards are easily understood to be an important ingredient in the new economy.

There is no evidence that the authors in this volume have ever addressed standardization, or indeed in the literature they conventionally cite. In part, this is because standardization almost never occurs at the level of the region, and many authors here have associated themselves with new regional models of economics. More generally speaking, standardization is an arcane activity, attracting almost no academic analysis from anyone other than a few economists. Involving a myriad of administrative and technical relationships, it seems an unlikely topic of interest for anyone whose primary goal is understanding changing economies or engaging in social theory. There are strong affinities between standardization and some of the themes addressed in this volume, however, especially concerning collaboration, institutions, social learning and new social relations.

Specifically, the essence of standardization is collaboration. Standardization fosters collaboration so that a common set of technical constraints for new technologies, networks, management systems and workplace practices can be developed and adopted. Further, standardization occurs in a complicated network of institutions. So complicated is this network that it is very difficult to identify the actual locus of decision-making. As well, in many ways, standardization involves 'learning institutions'. Its institutions are currently undergoing precisely the kinds of changes envisioned in this volume.[1] Their constitutions are being fundamentally rewritten. Committees are continually being struck to encourage institutional self-examination. New forms of organization are being created to take account of the particular demands of the new economy. And, finally, while no one would seriously claim that north–south relations, worker dislocation and unemployment are paramount in the list of concerns being addressed within standards institutions, neither are these issues totally off the agenda. Indeed, within the complex of standards institutions are several which deal directly with the second challenge noted above.

This case study of communication and information standards strains against the analysis presented in this volume by most of its authors, however, because it offers a fairly negative assessment of key elements of the new economy: collaboration, networking, learning institutions and so on. It also raises questions about the second challenge – that is, whether the new institutions can engender any democratic and humane response to the pressures introduced by the new economy. The goal of this paper, seen in the context of this volume, is to shed a different light on the notion of the new economy and its 'learning institutions' from that afforded by the other case studies. It needs to be said that much remains highly persuasive about the analysis elsewhere in this volume. If the case study of standards can make that analysis a little more robust in light of the many contradictions of the new economy, it will have served its purpose.

As just noted, not much is known about standards or the process by which they come into being. To understand how these most collaborative, networked and self-reflective of institutions might challenge the analysis offered in this volume, it will be important to understand how standardization actually works. For example, those engaged in standardization speak continually about collaboration, but it is essential to know what they mean by 'collaboration' before assessing the import of their collaboration in the new economy. Similarly, in the intricacies of standardization, one can learn much about the nature of networks, the meaning of flexible production, and the significance of rhetoric about the 'information poor' in the new economy, but only if one attends to the details of how decisions are actually made about standards. In the next section of this paper, then, some effort is made to introduce the reader to how standardization takes place. What follows sets the stage for discussions more germane to the subject of this volume.

Background

What is standardization?

Standards are nothing more or less than decisions about the way things will be produced or operated. They are decisions, usually reflected in numbers or technical documents, about the accepted or acceptable ways of doing things. Take the case of tea kettles, where the numerical standards reflect agreements about how hot kettles should get before they shut off. With hockey helmets, standards pertain to decisions about their acceptable resistance to impact. In the case of pollution, standards indicate what amount of negative effects can (or rather, will) be considered tolerable. There are standards for all of the production line equipment (setting minimal expectations for how it will perform) and for all of the attendant work, safety and industrial practices in industrialized production. In the new economy, standardization implies that one piece of equipment can usefully be connected to another to fashion a system or network, and/or that technical resources will be used in widely ('globally') accepted ways. In the new economy, there are new international standards for management and information systems and accounting. Indeed, in the new economy, as in the old, there are standards for virtually all aspects of industrial and financial activity.

Standards are not just related to administrative and technical matters, even if they look to be so from the outside. Take the example of the ISO 9000 series standards, which establish common management practices. These standards effectively separate large firms, that is, those capable of paying for and meeting these standards (and of being certified), from smaller firms, which lack the resources to do so. The ISO 9000 series is thus more than a badge of honor. It has quickly become a reliable guide to the serious players in the new economy, and a barrier to entry for the less well endowed. Similarly, standards for information network capacity and design determine *whose* equipment can, and will be able to connect with the network. Standards for software determine *whose* products will be attractive (because of compatibility) to users. Having standards for the massive information systems in the financial and banking services has resulted in ATM machines and geographically independent production, both of which are not inconsequential in producing worker dislocation. The Internet is a collection of standards (and technical protocols, which are similar to standards) which allow for new uses of existing communication networks.

The process of standardization

To understand something about the many, diverse and interconnected institutions of standardization, it is important, first, to know that standards can be developed in any number of ways. A single firm can have its own standards, for example. If these standards are proprietary, they act as barriers to competition in the sense that they ensure that competitors cannot connect with, or manufacture alternatives to, what has been put in place by the firm involved. Of course, unique standards are effective only if the firm involved has sufficient market clout (or regulatory protection) to maintain its exclusive purchase on the necessary standards. A small firm developing unique standards for its own products is unlikely to find clients, because prospective clients want their purchases to be compatible with what they already own. Only the largest firms can get away with having unique standards in today's markets. But communications is a domain where very large firms do exist, and their standards are quite effective in retarding competition.[2]

Alternatively a firm with market dominance can seek to impose its own standards upon everyone in the marketplace. If it is successful, its standards become de facto standards, and are used as if they were 'off the shelf' or neutral products. For competitors, adherence to de facto standards is required if they are to sell their equipment because, otherwise, technical incompatibility will render their non-standardized products, services or networks unattractive or even useless to customers.[3] Coercion is very much in evidence with de facto standards, in other words, but this is coercion delivered through market mechanisms. An example will be useful. Microsoft has mainly followed a strategy based on de facto standards (although an aspect of the firm's current legal troubles pertains to its allegedly covert use of company standards as well). Microsoft standards have become the standards for almost everyone. In this instance, all other firms are under considerable pressure to produce equipment and software that can be used in conjunction with Microsoft products. Even governments are under pressure to declare Microsoft as 'the standard' for the purposes of their very large procurement.

At first glance, it seems untenable to associate the kind of standards just discussed, company and de facto standards, with the notion of social learning or institutions, but this is so only if 'institution' is understood to be synonymous with organizations. There is another sense in which 'institution' is used in the literature. In this second use, 'institution' is more properly synonymous with social relations. It refers to the structural forces and constraints shaping crucial relationships in the economy. Seen from this perspective, both firm and de facto standards are 'institutional' because they set into place a system of relationships differentiating, for example, frontline participants from others in the new economy. To be in a position to institute company or de facto standards is to be at the front line. To enforce those standards upon others, through procurement policies or even by lobbying them into government regulations, is to position others in the back row, constraining their capacity to act or innovate.

There are also formal standards organizations. ISO, which developed the ISO 9000 series, is a good example, but there are literally hundreds of standards organizations operating at national, 'multinational',[4] and international levels. For the sake of convenience, I will call the company and de facto standards just discussed 'informal standards' and the kind of standards set by organizations such as ISO 'formal standards'. In theory, the formal standards organizations do not deal with informal standards. In practice, there is continual interaction between formal and informal standardization. This is so because the formal standards organizations are entirely dependent upon contributions from their members or participants, and these participants are firms also engaged in informal standardization. These companies also pay for the formal standardization meetings, the costs of their own officials to travel to meetings, and the expert reports relied upon in formal standardization. Companies develop informal standards even while they provide the technical data about products, processes or systems to serve as the basis for formal standards. Even as they produce and use informal standards, they engage in elaborate processes for approval and the publication of formal standards. They even buy the formal standards at tremendous cost to themselves. They do so even if their own engineers have developed the formal standard (and rendered it congenial with their decisions made about informal standards) and thus have no need of the information it contains. That formal standards often reflect (or officially sanction) informal standards should come as no surprise as a consequence. Even though, by rule, competitors must be involved in formal standardization, and though compromise and consensus are the name of the game, at the very least, the deliberations never begin with a blank slate. Informal standards are always taken into account.

This interchange between formal and informal standards is further supported by 'feeder organizations'.[5] Many 'feeder organizations' are actually industry trade associations whose members are active in the formal standards organizations (even while creating their own informal standards). The European Computer Manufacturers Association is a good example of a 'feeder organization'. 'Feeder organizations' do preliminary work to arrive at standards, but usually do not themselves create formal standards, but simply issue recommendations. In recent years, this role of the 'feeder organizations' has been explicitly recognized by several of the formal standards organizations, which now see 'feeder organizations' as the essential first stage, and themselves as operating primarily only at the last stage in standardization. In attempting to reform the standards process, several formal standards organizations now see their main role as authorizing and disseminating standards developed elsewhere, either as firm-based or de facto standards and/or within the 'feeder organizations'.

Finally, no description of standardization would be complete today without including the new forms of organization which have sprung up, in part to deal directly with the pressures of the new economy. These are strategic alliances, consortia and forums. In the case of strategic alliances and consortia, standards come into play because what is being held as 'co-operative' in these relationships is often an agreement about the technical parameters of innovation. Members of the alliance or consortium produce standards (and technical protocols) for their members as a byproduct of their activities.

Forums are somewhat different. With forums, standardization is often the point of the exercise. A forum is simply a conference, or series of meetings, organized around a particular technology, technical protocol or standard (set of standards). The forum constitutes a venue for explaining, and perhaps revising the standards. It serves as a place where suggestions for applications of standards can be made. Most importantly, the forum provides an opportunity for disseminating a standard (or set of standards) to other firms. Where there are competing standards in the market, forums are often organized by one or more of these competitors, each seeking to promote its own standard through its forum.

The institutions of standardization should thus be understood as including a diverse collection of organizations but, even more importantly, also a system of relationships (Salter, 1994). Included are individual companies acting in private capacities, companies seeking to achieve market dominance by imposing their own standards upon everyone, formal international organizations like ISO, intergovernmental organizations like the International Telecommunications Union, many other formal organizations operating at the national and multinational levels, 'feeder organizations' (some of which are industry trade associations), forums and strategic alliances. The processes for standardization are, in each case, often very different from each other. They include market relations, at one extreme, and highly deliberative formal processes involving governments, at the other. As different as they are, all these organizations are loosely interconnected, however. They are connected both through their members' participation in several of these organizations at once, and through the movement of recommendations from one organization to another. They are connected by the close relationship of formal and informal standards. Not only is there extensive overlapping membership in the 'feeder', forum and formal standards organizations, but all the major participants in the organizations are also actively promoting their interests through the marketplace, using informal standardization.[6] The system is, in short, a highly complex one. Nothing within it operates discretely; nothing operates exclusively according to its own rules. There are many instances of overlapping deliberations and even some examples of duplicate (triplicate, and so on) formal and informal standards for the very same technologies.

More needs to be said here about this system of relationships. Standardization always involves negotiation.[7] The rules for negotiation differ depending on where it takes place, however. This fact is used strategically by all participants. For example, representatives of a third world country would speak only in the plenipotentiary conferences of the UN-sponsored intergovernmental standards organizations, because in this venue alone are the rules structured in their favor. The representatives of IBM Europe, on the other hand, are likely be be found in virtually every important group where IBM interests might be impinged upon. IBM, of course, also acts through the market, relying upon its own standards and also trying to ensure that its own standards become de facto standards for everyone else – thus, like the third world representative, IBM 'forum-shops' to find the venues most congenial to its interests. But having a truly different position within the system, IBM has many more choices about how to achieve its goals than does the third world country. The rules for negotiation, both formal and informal, are conducive to IBM interests in multiple venues, while they are conducive to third world interests in only a few.

Choice is the watchword in this system of relationships, in fact. Choice abounds for a firm like IBM because IBM can choose different venues from time to time or from issue to issue. Thus, hypothetically, the representatives of IBM might well stay silent in the debates in the UN-sponsored agencies and, at the same time, be very active in some of the technical working groups and 'feeder organizations', but not others. Because the network of standards institutions is so loosely linked, IBM can easily make strategic decisions about which of the many formal standards organizations its representatives should frequent (and how extensively) by assessing the rules (and the relative clout of the other participants) all in order to determine how best to further IBM's interests. Opportunities for any large firm to choose how to operate within this system of relationships are almost limitless. Indeed, those who participate are not always seeking standardization, contrary to their rhetoric. Many times, firms participate in the various standards organizations simply to delay or prevent standards being produced.[8] For example, a firm would choose to delay standardization if it had its own products under development, and wanted to see the process delayed until its version could become the standard. A firm might reasonably want to prevent standardization if it was currently the beneficiary of market competition (recall that standards do retard competition in many cases). No participant ever admits to delaying or preventing standardization, but all participants in standardization understand that this often occurs. In short, participants can choose not only when and where to participate, but for what purpose.

There is also considerable latitude about the nature of the final product, the actual standards. A standard can be called a 'recommendation' or a standard. It can be called a 'guideline', or it can be issued as a 'technical report', or as an agreed-upon 'technical protocol'. 'Codes' are also standards. Standards can be developed by consensus (a 'consensus standard') and authorized and distributed by an accredited (private-sector) or intergovernmental organization. As such, it has special authority, even though it is voluntary. Or an industry trade association can develop a standard. Furthermore, any of these – voluntary and consensus standards, recommendations and guidelines, technical reports and protocols or codes – merely signal agreement. They are voluntary unless and until they are made coercive, for example if governments 'reference' them (adopt them, sometimes with revisions). And indeed, governments can themselves develop standards, which again can be voluntary ('often called criteria') or mandatory. In the European case, some standards straddle the line between mandatory and voluntary, inasmuch as they are approved as 'essential requirements' by the European Community, but have no import unless they are also adopted as regulations by national governments. In short, the notion of voluntary standards encompasses a whole range of different kinds of standards and, similarly, mandatory standards also vary with respect to their actual status, import and enforcement.

The point here is not to provide a comprehensive list of all the types of 'standards' (a task barely begun), or all types of standardization, but to illustrate the nature of the negotiations, which produce 'standards'. It is also to underline further how much choice there is for participants in standardization. To summarize: any large firm can choose when and where to participate, and it can make different choices from time to time or issue to issue. It can choose to further standardize or to delay or prevent it. It can operate mainly within the informal sphere or put emphasis on formal standardization. It can choose which kind of 'standard' it wants to see emerge from the process, and how forcefully to lobby for the resulting standards being mandatory for others as well. Finally, through informal lobbying, it can exert influence over how mandatory the so-called mandatory standards actually are. All of this occurs through a network of connected institutions, both formal and informal, each of which serves as a venue for negotiation, as do all of them together. The system of relationships sustains many kinds of interaction and interchange, even while it promotes choice for its more affluent participants.

With these characteristics of standardization in mind, it is now possible to turn to the central issues addressed in this paper.

Assessing the import of standardization

In the introduction, two challenges were identified that seem to preoccupy writers in this volume. The first was leveled at both firms and governments, and it involved creative adaptation within the new economy on its own terms. In this case, the focus was on institutional learning. Institutional learning can mean many things. It can involve collaboration, and the creation of new networks. It can mean both firms and governments leaving old patterns and rigid organizational structures behind and becoming responsive to the new economy. It can mean flexible production. As should now be evident, standardization involves collaboration, networking, responsiveness and even, in some senses, flexibility. Whether standardization manifests institutional learning remains to be seen. There was also a second challenge: creative (that is, democratic and humane) response to the pressures introduced by the new economy and its negative by-products. This second challenge will be taken up in the next section.

Collaboration

Each formal standards organization promotes and structures collaboration, and the system as a whole is premised on collaboration as well. Even unique company standards must achieve some acceptance and influence in the market, and among competitors, if they are to be useful, and this implies something akin to collaboration. That is, the very notion of standardization implies that decisions have been made, implicitly or explicitly, about what will be considered 'acceptable' to firms and governments whose interests otherwise diverge. Standardization would seem to meet one of the criteria for successful learning institutions in the new economy.

All is not what it seems, however. In fact, collaboration has many meanings in standardization. Only a few correspond to what is associated with institutional learning by authors in this volume. Indeed, to speak of standardization as being collaborative implies a level of commitment to standardization not much in evidence. Recall that, collaborative as they might be, participants in standardization move in and out of active involvement with various standards organizations as suits their purposes. They also often participate to prevent or delay standardization, in effect collaborating in the process only in order to avoid collaboration about the actual technologies. Despite all the so-called collaboration, there can exist several competing standards for the same technologies, because so little actual coordination is involved. Furthermore, it is commonplace to find that the mandates for the various working committees (even within a single organization) overlap with respect to some aspects of the particular technologies being considered. Thus, while it is true that many major firms invest huge resources in standardization and, in doing so, speak glowingly about the benefits of collaboration, true collaboration is much less common, even about relatively minor issues.

Despite these problems, the standards themselves seem to offer evidence of extensive collaboration. After all, as noted, standards are agreements among erstwhile competitors about characteristics of technology, products and processes. In some cases, these agreements also have the blessing of formal or intergovernmental organizations. Here is the problem. In arriving at a formal standard, those involved work exceptionally hard at achieving something akin to agreement, but nothing requires those who have negotiated the agreement to abide by the standards they themselves have created. There are many instances where standards are ignored. More commonly, participants write so many options into a formal standard as to make the standard virtually meaningless. Each firm can then claim that it is adhering to the standard even while it charts a different technological path from its competitors. It does so by following some options and not others or by actually relying on informal standards instead.[9] What appears to be a consensus, forged in collaboration, is actually an agreement to differ, to follow different courses of action based on the different options built into the standard or different options available to companies.

Furthermore, there is much room for slippage between those who fashion agreements about standards and those who make company decisions. The middle level engineers who negotiate standards into being are not the same people (nor are they closely linked to) those who do either research and development or those who do marketing in their own corporations. While none of these engineers would put forward proposals inconsistent with the technological strategies of their firms, neither do these engineers have enough clout to direct firm strategies by virtue of their actions in a standards organization. This means that, even if genuinely collaborative agreements were to be achieved within standards organizations, these agreements do not necessarily represent collaborative action on the part of the sponsoring firms. It would not be surprising to find, for example, that the senior management was more or less unaware of what the engineers had accomplished. Nor would such senior management be likely to refer to participation in standardization as reflecting the company's commitment to collaboration, or as involving legal or organizational collaboration with particular firms.

For the most part, the aspects of technology being standardized are fairly basic, and thus are far removed from the final products or systems being commercialized. Indeed, formal standards organizations now take great pains to ensure that they do not standardize items close to or in actual use commercially. Their leadership will speak about standards as being 'pre-competitive', and about limiting standardization to only those aspects of technology that really require it. They do not really believe that interconnection requires a single technological strategy, or that potentially incompatible technological decisions are counterproductive. In this sense, collaboration in standardization is limited to a very small component of the technological activities of any firm. Firms' decisions to collaborate more meaningfully occur in connection with R&D, ownership issues or on actual product development, and these decisions are taken in forums other than standardization.

One reason why it appears that standardization is so closely connected to collaboration, despite much evidence to the contrary, is that many standards decisions are taken by consensus. Nothing can move forward unless consensus is secured, and 'consensus' implies that everyone has worked together to achieve agreement. Consensus in standardization means something other than it does in everyday parlance, however. In the case of standards, it means that there are formal decision-making rules, that various interest groups are represented and that no votes will be taken. Practically speaking, it means that those who oppose (or are not very interested in) any standard can decide to remain silent about it, because they judge the standard to be irrelevant or insignificant to their own interests. Consensus means the absence of sustained opposition, as opposed to collaboration or an agreement among all present. Even in voting organizations, formal votes are discouraged, in order to permit 'consensus' to emerge despite opposition. There is, in other words, an unwritten rule that 'agreements' can be reached even where many are less than enthusiastic, and where others actually oppose what is being done (but find it unnecessary to do anything serious about it).[10]

If collaboration does not mean agreement, nor even sustained participation, what does it mean? First, collaboration is best understood as a commitment to keep many standards organizations alive and functioning, so that venues will exist for activities which participants find useful. The development of actual standards is only one among many such activities. These activities range from exchanging information and gathering technological intelligence to negotiating very basic agreements to support technology investment and securing access to scarce resources such as radio spectrum. They include looking for changes to public policy. Second, consensus means having well-defined decision-making rules and accepted conventions for these diverse interactions and negotiations so that potential participants will know in advance whether the rules privilege their interests in each venue, and whom else they are likely to encounter by participating. Without this information, they cannot take advantage of the choices discussed above, nor act strategically.

Third, collaboration is involved when firms create mechanisms for officially recognizing standards. Once officially recognized, standards appear to be agreements about the best possible technical practice. They appear to have no connection to particular firms, but instead to represent an overarching interest in achieving technical solutions to common problems. Officially sanctioned, standards are disseminated through arrangements with national standards bodies (including in third world countries) throughout the world. Collaboration in this instance means collaboration after the fact, and it occurs when firms or countries, which had no part in developing them, adopt the standards. Interestingly, it is often the least technically advanced countries and the smallest firms who collaborate thus, harmonizing their activities to what they believe to be a neutral product, the best solution to common problems.

Finally, collaboration means establishing some common ground-rules for those who can speak authoritatively about these best technical solutions to common problems. It means creating a common language to speak about both political and technological issues. Standards organizations are 'talking clubs', where participants set agendas for public policy, and define the terms for public debate. Their role in this regard is all too easily overlooked, because the link between cause (talk) and effect (different orientation to policy-making) is virtually impossible to track. What is apparent, however, is how much time (in recent years, an informed estimate would be 50 per cent or more) is spent in standards meetings in fashioning words, statements and approaches to technological development, words which support and legitimize liberalized, deregulatory policies on the part of governments. It is interesting to note that the participants claim standardization should be industry-led, and nonetheless they speak extensively about influencing government.

With all these points in mind, it is now possible to decode 'collaboration'. In the case of standardization, collaboration is only peripherally about an end product (the standards). Collaboration is best understood not as relating to standards, but as commitment to a process, as shaping a discourse, as a form of risk management, as legitimization, and as establishing venues for interaction. Ensuring that such venues exist, that new government policies are fostered, that the 'rules of the game' are known in advance and, consequently, that choices can be maximized in a relatively risk-free environment are matters that everyone agrees are important, even if they themselves choose not to participate in the actual standardization. Whether any of this deserves to be called institutional learning seems less likely.

Networks

At first glance, standardization seems not to be characterized by networks. Many standards organizations have no official relations with each other at all. Indeed, it is truly astonishing how little connection there appears to be (despite official protestations to the contrary) between organizations, or even technical working groups, working on the *same or related* technologies, this in spite of the fact that the organizations or committees have some of the same members and that much rhetoric is devoted to encouraging the opposite. Worth reiterating is the point made earlier that standardization facilitates a multiplicity of choices for each of its major participants. As such, it depends upon there being distinct venues, each having its own rules. As said earlier, nothing prevents a company like IBM from participating in several organizations at once (or not), and differently from time to time. Furthermore, many standards are informal, and thus subject to only as much networking and coordination as markets make possible.

Officially, the story is otherwise. There is supposed to be a chain of action linking national standards bodies and 'feeder organizations' with decision-making bodies at the 'multinational' and global levels. Participants speak glowingly about how the national and 'multinational' (for example, European) standards organizations exist only to facilitate the work of global standardization. They regularly pay tribute to the international standard organizations as being pre-eminent. But these same people also recognize that there will be areas of standardization, which are not (or not yet) amenable to global standards. And they understand that political and economic rivalries, between Europe and the United States, for example, will result in participants putting aside their supposed commitment to global standardization in favour of their own interests. Despite widespread agreement that standardization should be coordinated through a network of linked organizations, no one believes that coordination is actually commonplace or that it will seriously impinge upon any participants' freedom of action.

Yet there are connections, a network of sorts. As noted, the membership lists of various organizations overlap significantly with each other, such that many of the same participants appear at every important standards meeting. Even where there are no formal connections, the technical committees continually reference each other's work. Indeed, the first step in any discussion about a new standard will be to determine what other standards organizations have done. This will be seen not only as a point of departure, but as a serious constraint upon doing anything differently. Formal standards organizations also rely heavily upon 'user' and 'expert' committees, whose membership reads like a roster of the most influential firms worldwide. Here again, one finds the same names and firms involved in 'user' committees of different organizations.

In short, there is a contradictory nature to the networks manifest in standards. Even though people and organizations are only loosely interconnected, there are truly close linkages as well. A framework exists for knitting together interests, even though formal coordination is more rhetorical than real. This contradiction is not problematic for participants in standardization, however difficult it might be for those who want to describe the situation to outsiders. Insiders understand all of the connections, however tight or loose they may be. Participants know how 'to work the system' to their own advantage, maximizing their freedom of action at one moment, and using standards organizations to constrain everyone else at another. To speak of standardization as being characterized by networks is to stress the informal connections which make the freedom to make strategic choices possible. Whether this kind of networking furthers institutional learning is, at best, also an open question.

Responsiveness and capacity for change

In the past decade, standards organizations have demonstrated a great capacity for change. To be sure, there still exist organizations like the International Telecommunications Union, organizations, criticized severely for their resistance to change, the implacability of their senior staff, their increasing irrelevance to industry and the market, and so on. But even in the ITU, and certainly in other formal standards organizations, a great deal of time has now been spent on redesigning the organization, changing the membership, reforming decision-making practices – in short, on changing how standardization is carried out to meet the demands of the new economy.[11] Indeed, so prevalent is the reform in many standards organizations that those who actually sit in the technical working committees (which actually develop standards) have complained that their work is being ignored.

The results of all this responsiveness are easy to see. Standardization has been speeded up dramatically. The work of the forums, strategic alliances and 'feeder organizations' has been acknowledged, and these groups are no longer seen to be competitors with the formal organizations. Less is now being done to create standards for everything and, instead, priorities are being set to focus attention on aspects of technologies truly requiring standards. These are not insignificant changes, yet here too there are contradictions. For example, one would think that with so much soul searching (and with an increasing shortage of money to support standardization), some of the old-style standards organizations would have gone out of business. By the same logic, efforts should be underway to merge organizations or at least working committees, and to share work. Not only is this not happening, but new organizations, committees, working groups, forums and so on, are being created all the time. And, where work is supposed to be shared, the usual response is to create yet another new committee, and to have this new committee report back to each of the older organizations. To be sure, many of the newer groups look nothing like the old formal standards organizations and many are also task-driven and short-lived. That said, the organizational landscape is more crowded and more diverse than it was even five years ago. If responsiveness were truly driving the reforms to standardization, one would expect the opposite.

Recall, however, that participation in each organization (or committee) supports different strategic choices, and choice is the watchword of the system. Failing to find a venue conducive to its interests, it would not be surprising if a firm were to launch a new initiative, propose a new joint committee, or support an ageing organization which otherwise had outlived its purpose in order to secure a more congenial venue for advancing its interests. Given this, the number of venues might actually be expected to increase, not decrease, despite all the streamlining.

Furthermore, to an outside observer, it seems like an inordinate amount of time is being spent on restructuring. In many organizations, immediately following a lengthy negotiation about how restructuring will be done, the process begins anew. Why are so many organizational (and firm) resources being expended on what appear to be matters of housekeeping? And why, like housekeeping, is reform never-ending? The answer is simple. Strategically, it is useful to keep the rules in flux because participants are always trying to reshape them to suit their needs.

The notion of responsiveness implies that firms and governments are going through a process of learning, moving from old-style approaches to practices more congenial to the new economy. What is occurring in standardization looks to be responsive to the dictates of the new economy, but it can hardly be said to constitute institutional learning. Continual negotiations to further strategic interests may well be characteristic of the new economy, but the organizations within which they take place are not really learning institutions, nor is there as much streamlining as first appears to be the case.

Flexibility

To many, flexibility means downsizing, corporate and capital mobility without constraint, lack of regulations or liberalized trade regimes. Almost all of the participants in standardization endorse these notions of flexibility. There is another sense in which the term is sometimes used, however, one which links flexibility to production. In this second sense, flexibility refers to the reshaping of the production process away from the old model of mass production towards 'just-in-time' production or something similar. There is some evidence that standardization is becoming more flexible in this second sense, assuming that standards are 'produced' like any other product (they are sold as a product).

In the past, the formal standards organizations amassed their resources through membership fees and the sale of their standards. They were able to carry out extensive work programs because their participants (many of whom were government officials) paid their own travel and meeting costs. Because technical committees determined their own work plans and because such committees were staffed by engineers, the agenda for standardization encompassed almost all aspects of the technology. That relatively few standards were actually produced, and that it took many years for any single standard to emerge, seemed not to be a problem, because the pace of technological development was relatively slow. All that has now changed, of course. As noted, standards organizations today focus on relatively few aspects of technology. They rely upon 'expert' and 'user' committees extensively to help set priorities. The time required to develop any standard has been dramatically curtailed, and there are fewer stages in the approval process. Almost all organizations have developed a 'fast track', for some standards. Surveys have been conducted to gain information on what firms need, and about how firms feel about standardization. In such cases, work will not be authorized unless those who intend to use the standard provide resources for its development. In short, something akin to 'just in time' production has developed.

There is a contradiction here too, however. It does not lie with standardization so much as with the influence of the newly flexible standards organizations. Put simply, the point is that less reliance is being placed on these newly flexible organizations to develop standards. Recall that at the beginning of this chapter it was noted that formal standards organizations comprise only a portion of standardization. It was noted that firm-based and de facto standardization were important components of standardization, and that the informal mechanisms of the market were equally capable of generating standards. Recall also that it was argued that forums had become an essential component of standardization. Forums are intended not to set standards so much as to promote applications of particular technologies (based on standards) or to market particular standards, often informal standards. What has happened instead is that informal methods are becoming the main route for standardization. Notwithstanding how much effort formal standards organizations expend becoming compatible with the new economy (adopting flexible methods of production for themselves), they are becoming irrelevant (Cargill, 1997; Bahke, 1998). Learning institutions though they be, they are being replaced within the new economy.

Concluding remarks

If indeed, standardization is intimately connected to the new economy, this case study offers some sobering reflections on what it might take to meet its challenges successfully. To be sure, standardization is nothing if not collaborative, involves many informal networks (and much rhetoric about the importance of networks), and has many examples of 'learning institutions'. In the production of standards there is also evidence of something akin to just-in-time approaches. But the analysis offered here has suggested that all is not what it seems. Collaboration means something other than cooperative relationships or agreements in the case of standardization. Networks do exist (as evidenced by overlapping memberships in the various kinds of standards organizations) but networks seem more to promote advantage seeking on the part of individual firms, facilitate forum shopping, and expand choices for individual firms. Even while formal standards organizations are becoming responsive to the new economy, and more streamlined and flexible in their production of standards, they are becoming less influential. Informal standardization is rapidly overtaking formal standardization.

To make this argument does not call into question the close association of learning institutions (collaboration, networking, and so on) and the new economy. The changes occurring in standardization are, without doubt, manifestations of adaptation to the new economy. Terms like these have a nice ring to them. They imply that somehow firms are moving away from the destructive aspects of capitalism, based on uncompromising competition, towards a more cooperative approach. They imply that social benefits will flow from the new economy, as well as economic ones for its most influential participants. What is called into question by this case study is what these terms imply. Just as the old order might not have been as fiercely competitive as it seemed, so too the new economy might be considerably less cooperative than it seems, despite all the collaboration and networking. The social benefits flowing from responsiveness and flexibility might well be much more limited than is implied by many writers on the topic. In pursuing the line of thinking evident in this volume, it might have been better simply to describe, in ever greater detail and with ever more nuance, such developments, leaving aside the question of whether these developments are progressive for anyone other than the most influential participants in the new economy (Mansell, 1994; Huggins, 1997; Casson, 1995).

Recall however that there were two challenges posed for the new economy. The second has attracted far less optimism even from those who write prescriptively about how to adapt to the new economy. Almost everyone agrees that generating a creative response to the dislocations and other negative effects of the new economy is a tough job, one not likely to succeed without a great deal of attention and political pressure being applied.

Dealing with the second challenge

Those who set standards are not without awareness of social issues, even if they are mainly engineers within middle management of large firms. For example, they deal regularly with what they call 'human interface' issues. Even in the communication and information technologies, consumer and worker safety issues often arise in connection with standards.

Standardization is understood by all its participants as both a management and a technical activity. In their view, expertise separates real participants from those who should not really be taken seriously. Though much of what is being accomplished is closer to policy-making than technical work, technically expert individuals, companies and countries are privileged. In dealing with human interface issues, neither labour nor consumers has much presence or credibility in a managerial or technical discourse. And, while social policy issues, such as worker dislocation, are sometimes raised, it would be hard to find any evidence that standards decisions had been affected as a result. In short, concern about social issues is filtered through a technical and management discourse, so much so that it often appears to be absent.

The exception occurs in connection with what might be called resource issues, especially in connection with developing countries. All participants in standardization believe that the gap between the 'information rich' and the 'information poor' is increasing, and all believe that something should be done about it, although obviously, private sector organizations and firms prefer to leave the task to public organizations and governments. Organizationally, only the intergovernmental organizations are structured to permit the 'information poor' to have a say in standardization, at the general assemblies which, theoretically, are the supreme decision-making forums of these organizations. Not surprisingly, however, few in developing (or poor) countries have the necessary technical expertise to speak credibly about standards. They also lack the financial resources to attend the myriad of meetings, forums, conferences and working groups where the main work of standardization takes place. Indeed, in recent years, even in the intergovernmental organizations, development issues have been distinguished from standardization, and separate 'bureaus' have been established to deal with them. The private sector organizations and firms similarly separate what they do about standards from their expressed concern about the 'information rich and poor'. The assumption is that the problem to be addressed is only how developing countries can gain access to new technologies (especially because many lack even extensive telephone service). The social task to be done is to make standards and technical guidelines available, so that developing countries can partake of some of the benefits delivered by the new economy.

This assumption has become controversial in recent years, however. No longer is it presumed that developing countries will follow an orderly progression starting from basic telephone service, culminating eventually in use of advanced technologies. This notion of progress now seems hopelessly naive. Rather, it is now thought that a small group within even the poorest of countries will be able to leap-frog the older pattern of technological development, moving directly to the most advanced communications and information systems. Multinational firms now bid competitively to serve the needs of these select groups, to install communication and information technologies in countries with little else to support the use of advanced technologies. It is understood that most of the population in the poorest countries will not benefit directly from such installations, although it is believed that economic development will eventually follow.

In this context, long-standing attempts by standards organizations (especially intergovernmental ones) to educate third world countries about basic communication technologies are now accompanied (if not supplanted by) demands from their most influential participants that standards should be set by the technically expert. Less and less is being said about the important role of the general assemblies of the intergovernmental organizations in disseminating technical information to help poorer countries catch up. Instead, participants openly advocate removing all the key decisions about standards (and the final approval for individual standards) from the general assemblies. Fear is driving this change in attitude about the role that standards organizations (especially intergovernmental ones) can and should play in addressing the gap between 'information rich and poor'. It is now believed that, if these organizations choose one standard and disseminate information about it, the interests of firms seeking to serve the same market with another technology or standard will be infringed. In short, the developing countries are now seen as important markets, despite their poverty. However real it may be, concern for raising the level of technology in poorer countries and closing the gap between 'information rich and poor' has been eclipsed by the drive to conquer these new markets situated within (but hardly an integral part of) these same countries.[12]

Because intergovernmental organizations are often involved, there appears to be a small provision for democratic response in standardization nonetheless.

This too is misleading. As noted, the intergovernmental organizations are structured in such a way as to keep both technical issues and much of the reform initiative in the hands of the experts. Even these organizations are fundamentally dependent upon their most influential corporate 'members' to have any data to support standardization. They have now set themselves up to take advice from their most influential participants. But what about at the national level? Surely in the national standards organizations, there is more room for a democratic response, because labour and consumer groups are often among the participants. This too is misleading, for two reasons. First, the national standards organizations today rarely originate standards, particularly in the field of communication and information technologies. More commonly they review, adapt and disseminate standards developed elsewhere, in the industry trade groups, private sector and/or international organizations. Despite their 'balanced membership' including labour, consumer and government representatives, and despite their efforts at public consultation, these organizations are no less dependent upon contributions and data from their industrial members than are other standards organizations. Even if it were true that these national organizations provided space for public discourse, it would matter little, because most standardization occurs elsewhere.

Almost all standards organizations, including the industry trade associations, the 'feeder organizations' and forums, regard themselves as being public, however. In this case, public and democracy are different. In the case of standards, 'public' means non-profit. Being 'public' means having meetings open to the public (often at a cost) or deliberations published. 'Public' also means that anyone who meets the membership criteria (including often a healthy membership fee) can join. As well, in standardization 'public' should be understood in connection with public service. The engineers who develop the standards, and work assiduously at reforming standardization, think of themselves as performing a public service in arriving at decisions to support innovation and development in the communications and information technologies. Their jobs notwithstanding, some of them are progressively minded people, with strong commitments to democracy, internationalism, and even, in a few cases, very humane social welfare policies.

Just as collaboration had a specific meaning in the domain of standardization, so too does the response to social and democratic issues raised by the new economy. As just argued, it is not that social issues are absent from discussions about standards, or that no one has ever cared much about the gap between the 'information rich and poor', or that no opportunities exist for democratic participation in standardization. Rather, social issues are dealt with as management problems. Concern for the gap between 'information rich and poor' has been separated from the main work of standardization, and eclipsed by the agendas of the most influential of its participants. There are opportunities for democratic participation, but they are extremely limited. What is meant by 'public' is something else entirely.

With respect to the second challenge described in the introduction to this paper, therefore, this case study again offers a few sobering thoughts. It would be wrong to suggest that the new economy (as manifest in standardization for communication and information technologies) is simply a closed shop, or that its participants have no concern whatsoever for the impact of the new economy (and of their own decisions) upon those disadvantaged by the new economy (indeed also within the old one). Those engaged in standardization are no different from those writing in the academic literature. Some are genuinely committed to fostering a democratic and humane response to the new economy and others care less. The point to be made is a simple one. Within the domain of standardization, and probably more generally in the new economy as well, these kinds of concerns are reconfigured so that the democratic and humane response no longer corresponds to what people usually mean when they use this terminology. Increasingly a gap exists between the meaning of these words in the specialized domains of the new economy and their meaning is everyday life. This fact too must be factored into any discussion of 'learning institutions', and into policy prescriptions for adapting to the new economy.

Notes

1 For the debate about the role of nation-states versus their regions, *see* Waters (1995); Mansell (1994); Huggins (1997). *See* Pospischil (1993); on changes in the ITU, *see* Besen and Farrell (1991); W. J. Drake presents an overview of changes in the world of telecommunication standards since the 1850s in 'The Transformation of International Telecommunications Standardization: European and Global Dimensions', in Steinfeld, Bauer and Caby (eds)(1994). *See also* materials from standardization organizations themselves – for example, the European Telecommuncations Informatics Services (ETIS), an organization of European telecommunications operators, struck a task force to guide ETIS through the change process, and at its 1997 Conference in The Hague sponsored a workshop entitled 'Addressing Complexity and Change' (1996). In order to better position itself in the newly competitive world of ICT, ETSI engaged in a 're-engineering' process in 1995, which resulted in a restructured organization (ETSI, 1996). One aim of both organizations as they restructure is to enable better cooperation among ICT parties. Richard Hawkins (1995) questions whether the outer organizational changes have produced matching inner changes in a paper presented at the OECD Workshop of ICT Standardization in the New Global Context, Paris 1995.

2 Participation in the standards process in general is biased even against SMEs because of the costs involved: *see* Drake, (1994); David and Steinmueller (1996). Carl F. Cargill gives the example of the demise of a small British tripe manufacturer who had to close down after new EU standards for food-processing plants came into effect (Cargill, 1997). But, however difficult it is as an SME to participate in the standardization process or obtain process certification, SMEs are among those

obtaining ISO 9000 registration. A survey of British-registered mechanical engineering manufacturing companies found that SMEs in particular benefit from it,
with small firms showing nearly three times the profitability of the industry
average, and better performance in other financial and sales measurements (ISO,
1996).

3   For a discussion of de facto compatibility standards and competition, *see* David
and Steinmueller (1996); also Besen and Farrell (1994); Hegert (1987); David
(1987); and Mansell (1993).

4   The proper term is 'regional'. Regional standards bodies bring together participants
from standards organizations within a region, such as Europe or the Americas. I
have used 'multinational' here instead of regional, because 'regional' has quite a different meaning in this volume. 'Multinational' standards bodies should be distinguished from international bodies, because they are restricted to particular countries.

5   Feeder organizations can be closed groups working within the formal standards
organization: *see* Kjell Strandberg (1998); on the role of the feeder organizations
for ETSI, 1997 interviews with Stienstra of Philips. ECMA tailors its projects so that
its standards can easily be taken up by JTC-1, and is also a feeder organization for
ETSI; 1997 interviews with De Ruyter van Steveninck of Philips and Jan van den
Beld of ECMA.

6   *See* Hurd (1998). On the differing clout of various firms in standards organizations, *see* for example ETSI's weighted voting system based on the yearly turnover
(and corresponding membership fee) of its member organizations. The need to be
involved in a variety of groups was discussed in the 1997 interview with Hartmann
of Siemens; *see also* R. W. Hawkins (1993). On standardization as a negotiation, *see*
J. Farrell and G. Saloner (1988).

7   *See* Hawkins, (1996); Atkinson, (1993); Coleman and Skogstad, (1990).

8   A classic example of compliance with an opposition to standards is in the
American automobile industry; *see* Thompson, 1954. For obstruction within ETSI,
see Hawkins, 1993; that standards are not always desired, an anonymous interview
(1997).

9   On ignoring or not implementing standards, *see* Hawkins (1993) and (1996); also
P. A. David and W. Steinmueller (1996).

10 *See* Cargill (1997). The new approach to consensus and voting in ETSI is discussed
briefly in Kjell Strandberg (1998).

11 The ITU is at the forefront of SDOs addressing social and north–south concerns. It
maintains a separate bureau for the advancement of telecommunications in the
developing world, and its inclusive membership ensures that even the least powerful
have some influence (Drake, 1994; Besen and Farrell, 1991). At the 1998 ITU
Plenipotentiary, it was agreed to forgo the interest on the outstanding contributions
of several developing nations (for example, Docs 144 and 145). Within the organization, provision of occupational health and unemployment benefits for its staff is the
topic of lengthy discussion, and the 1998 Plenipotentiary unanimously endorsed
the inclusion of a gender perspective within the ITU's work (Doc 147 and Corr 1).

12 Policy statements of numerous LDCs such as Burkina Faso, Djibouti and Tuvalu,
delivered at the 1998 ITU Plenipotentiary, make mention of restructuring and
deregulating their telecommunications institutions in order to allow multinational
competition and participation in ICT deployment; frequently wireless installations are considered of greater value than wireline in order to increase penetration
of services into the countryside.

References

Atkinson, M.M. (1993) 'Public Policy and the New Institutionalism', in M.M. Atkinson (ed.), *Governing Canada: Institutions and Public Policy*, Toronto: Harcourt Brace Jovanovich Canada Inc.

Bahke, Torsten (1998) 'The Future Role of National Standards Organizations', Proceedings of the Cen/Cenelec/ETSI Conference, Making Standards for the Market, Nice.

Besen, S.M. and J. Farrell (1991) 'The Role of ITU in Standardization: Pre-eminence, Impotence or Rubber Stamp?', *Telecommunications Policy*, vol. 15, pp. 311–21.

Besen, S. M. and J. Farrell (1994) 'Choosing How to Compete: Strategies and Tactics in Standardization', *Journal of Economic Perspectives*, vol. 8 (2), Spring, pp. 117–31.

Cargill, Carl F. (1997) *Open Systems Standardization: the Business Approach*, Upper Saddle River, NJ: Prentice Hall PTR.

Casson, M. (1995) *The Organization of International Business: Studies in the Economics of Trust*, Aldershot: Edward Elgar.

Coleman, W.D. and G. Skogstad (eds) (1990) *Policy Communities and Public Policy in Canada: a Structural Approach*, Mississauga: Copp Clark Pitman Ltd.

David, P. A. (1987) 'Some New Standards for the Economics of Standardization in the Information Age', in Partha Dasgupta and Paul Stoneman (eds) *Economic Policy and Technological Performance*, Cambridge and New York: Cambridge University Press, pp. 206–39.

David, P. A. and W.E. Steinmueller (1996) 'Standards, Trade and Competition in the Emerging Global Information Infrastructure Environment', Background Paper prepared for the ESRC/GEI Workshop on Competition, Regulation, Standards and Trade Policy for Telecommunication Services, London.

Drake, W.J. (1994) 'The Transformation of International Telecommunications Standardization: European and Global Dimensions', in C. Steinfeld, J.M. Bauer, and L. Caby (eds), *Telecommunications in Transition: Policies, Services, and Technologies in the European Community*, Thousand Oaks, CA: Sage.

European Telecommunications Informatics Services (ETIS) (1997) *Annual Report 1996/1997*, Brussels: ETIS.

European Telecommunications Standards Institute (ETSI) (1996) *Annual Report*, Sophia Antipolis.

Farrell, J. and G. Saloner (1986) 'Installed Base and Compatibility: Innovation and Predation', *American Economic Review*, 76, pp. 940, 955.

Harvey, D. (1989) *The Condition of Postmodernity*, Oxford: Basil Blackwell.

Hawkins, R.W. (1993) 'Changing Expectations: Voluntary Standards and the Regulation of European Telecommunication', *Communications and Strategies*, vol. 11 (3), pp. 53–85.

Hawkins, Richard (1995) 'Efficiency and Responsiveness in Standardization Organizations: Determining a Realistic Basis for Evaluation', paper presented at the OECD Workshop of ICT Standardization in the New Global Context, Paris.

—— (1996) 'Standards for Communication Technologies: Negotiating Institutional Biases in Network Design', in R. Mansell and R. Silverstone (eds) *Communication by Design: the Politics of Information and Communication Technologies*, Oxford: Oxford University Press.

Hegert, M. (1987) 'Technical Standards and Competition in the Microcomputer Industry, in H. Landis Gabel (ed.), *Product Standardization and Competitive Strategy*, Amsterdam: North Holland.

Huggins, R. (1997) 'Competitiveness and the Global Region: the Role of Networking', in J. Simmie (ed.), *Innovation, Networks and Learning Regions?* London and Bristol, PA: Jessica Kingsley Publishers and the Regional Studies Association, pp. 101–23.

Hurd, John. (1998) 'Why Does Digital Participate in Standards?', *Computer*, vol. 31 (6), June, pp. 95–6.

ISO (1996) *ISO 9000 News*, vol. 5 (1), January/February, pp. 8–10.

Mansell, R. (1993) *The New Telecommunications: a Political Economy of Network Evolution*, Thousand Oaks, CA: Sage.

—— (1994) 'European Telecommunication, Multinational Enterprises, and the Implication of "Globalization"', *International Journal of Political Economy*, vol. 23, Winter, pp. 83–104.

Pospischil, Rudolf (1993) 'Reorganization of European Telecommunications: the Cases of British Telecom, France Télécom and Deutsche Telekom, *Telecommunications Policy*, vol. 17 (8), pp. 603–21.

Salter, Liora (1994) 'The Housework of Capitalism: Standardization in the Information and Common Technology Sector, *International Journal of Political Economy*, 23 (4) Winter 1993/94, pp. 105–31.

Strandberg, Kjell (1998) 'Collaborative Standardization – a Vision', Proceedings of the Cen/Cenelec/ETSI Conference, Making Standards for the Market, Nice.

Thompson, G.V. (1954) 'Intercompany Technical Standardization in the Early American Automobile Industry', *The Journal of Economic History*, vol. 1 (1), Winter, pp. 1–20.

Waters, Malcolm (1995) *Globalization*, London and New York: Routledge.

5
Locational Tournaments, Strategic Partnerships and the State

Lynn K. Mytelka

Introduction

There is widespread agreement in the literature that over the past two decades, the knowledge-intensity of production has increased across a broad spectrum of industries and competition has both globalized and become more innovation-based.[1] A consensus also seems to be emerging that to survive in this new competitive world, firms, regions and states must become part of a 'learning economy' in which learning and unlearning of habits and practices, institutions and conventions is a continuous process.[2]

In a recent paper, Bengt-Åke Lundvall (1995) observed that traditional economic theory, which is concerned with the making of choices between well-defined alternatives and the allocation of scarce resources, deals less well with the uncertainties and dynamics that are integral to the learning economy.[3] For Lundvall, adopting a learning and innovation perspective will thus require a shift in focus away from the allocation of existing resources in the context of a stable set of parameters, to the creation of new resources in a situation marked by continuous changes in technologies, preferences and institutions. Under such conditions, this paper argues, 'learning to allocate' is less important than 'learning to learn' and the future position of firms, regions and states will likely reflect their capacity to learn and unlearn.

This chapter looks at locational tournaments[4] and strategic partnerships from the perspective of learning and innovation. Its purpose is to explore the effects of turbulence created by globalized innovation-based competition and the heightened capital mobility to which it has given rise, on the learning environment for small and medium-sized enterprises (SMEs) many of which are suppliers in what have become international production networks. It thus situates these firms in two potential learning contexts, those produced regionally, through the impact of locational tournaments on local clusters of enterprises, and those made possible through spatially dispersed strategic partnerships. The focus is primarily on Europe, but in looking specifically at changing competitive conditions in industries such as electronics and automobiles it confronts head-on the phenomenon of globalization and its impact on regional economies elsewhere as well.

The first section sets out a framework for analyzing the relationship between the turbulence created by accelerated capital mobility and learning and innovation within local economies. The second section examines the potential for learning and innovation by SMEs through long-distance partnering stimulated by the European Union's research and technology development (RTD) programs in R&D-intensive industries, such as electronics, and through participation in more traditional local supplier–client relationships in the automobile industry. To set the European case more broadly, the paper makes a brief reference to the impact of foreign direct investment on the Brazilian automobile industry in recent years. The third section concludes that globalization is having a profound effect on the durability of local linkages. In some industries it also appears to be reshaping the possibilities for learning and innovation through long-distance partnerships. For the SME sector, however, it is clear that dynamic regional economies remain of critical importance in creating the capacity and accelerating the speed with which small and medium-sized enterprises are able to adjust to change. Policies to attract foreign investment must be rethought from this perspective.

Capital mobility and turbulent learning environments

Much of the contemporary literature on globalization regards capital mobility as inevitable or treats it as a social 'good', stressing the efficient use of capital that is assumed to result and ignoring the disruptive effects on employment and local development that it produces.[5] By the mid-1990s, estimates placed the value of all capital account transactions at between $900 billion and $1 trillion per day or roughly $350 trillion per year (UNDP, 1994, p. 69). The value of world trade in the year 1992, however, barely reached $5 trillion – $3,640 billion in world merchandise trade and a further $1000 billion in world trade in commercial services (GATT, 1993, p. 1). Thus, neither the needs of world trade nor foreign direct investment flows, which averaged only $140 billion per year over the period 1987–91 (UNCTAD, 1993, pp. 243–7), can account for the magnitude of capital account transactions. As much as 70 per cent of capital account transactions, therefore, are short term in nature, less than a year to maturity and most of this is probably speculative – arbitrages seeking to profit from exchange rate fluctuations or interest rate differentials. This was the background to the Asian financial crisis of the late 1990s and in retrospect it appears to have contributed to the banking crisis in the United States in the 1980s and the 1992 crisis in the European Monetary System (UNCTAD, 1998a, pp. 53–79). On several occasions, Nobel Laureate James Tobin has called attention to the fact that such capital flows

. . . contribute little to rational long-term investment allocations. Exchange rates are at the mercy of the opinions of private speculators commanding vast sums. Their activities distort the signals exchange markets give for long-range investments and for trade (UNDP, 1994, p. 70).[6]

Policy-makers, concerned about the effects of exchange rate fluctuations, have thus been induced to adjust domestic interest rates, while firms whose assets are affected by the volatility of exchange rates and interest rates shorten time horizons for investment decisions and increase their 'preferences for liquid as opposed to longer-term financial instruments' (UNCTAD, 1994, p. 95).

The increasing volatility of speculative capital has raised concern '. . . about issues such as the sustainability, composition and terms of capital flows, and the need to ensure that they are consistent with macroeconomic stability, international competitiveness, growth and social equity' (Devlin *et al.*, 1994, p. 2).[7] In the wake of the Asian financial crisis, the volatility of portfolio investment has come under particular scrutiny (UNCTAD, 1998b, pp. 13–19). Far less attention, however, has been paid to the acceleration in flows of what might be called productive capital and few such studies have assessed the impact of rising capital flows on the development of local systems of innovation and production. To the contrary, studies by the World Bank, IMF and UNCTAD have underlined the relative stability of foreign direct investment (FDI) and its non-debt creating nature. This message is not lost on national and regional policy-makers.

Thus as markets and competition globalized and direct foreign investment flows rose dramatically over the 1980s and 1990s, regional and national authorities found themselves increasingly under pressure to liberalize capital exports and enhance their attractiveness as sites for international capital. Remaining capital controls were lifted, nominal exchange rates were stabilized, and subsidies and other inducements were offered to influence locational decisions. This was accompanied by reductions in tariff and some non-tariff barriers to trade. With liberalization, an innovation-based mode of competition has generalized, making it increasingly more difficult for small and medium-sized enterprises to compete on their own. Pressures for further liberalization were maintained despite the Asian crisis and its ripple effect in markets around the world and by 1999 momentum had built for a Millennium Round of trade negotiations within the World Trade Organization (WTO).

As the growing knowledge-intensity of production led to higher R&D costs and shortened product life cycles, amortization of these costs over wider geographical markets became imperative (Mytelka, 1983, 1987). Liberalization thus induced a further acceleration in capital mobility and intensified mergers and acquisitions (M&As) in the world's largest markets – the EU, the US and a number of developing countries such as Brazil.[8] M&As currently account for well over 50 per cent of total foreign direct investment on a world scale (UNCTAD, 1999).

From a dynamic innovation perspective it might be hypothesized that accelerated capital mobility carries with it both new opportunities for learning and the threat of increased turbulence that makes trust-based networking and innovation stimulated by longer-term interaction more difficult. Elsewhere I have attempted to measure these effects in aggregate fashion by looking at changes in average annual inflows and outflows of foreign direct investment as a share of gross fixed capital formation and then relating these to changes in resident patent applications. The study showed that the rapid rise in the share of gross fixed capital formation accounted for by FDI inflows over the 1980s and 1990s was associated with a decline in resident patent applications in the United Kingdom, Netherlands, Denmark, Belgium and Sweden (Mytelka, 1999). To the extent that the competitiveness of individual firms requires a wide array of domestic linkages between users and producers and between the knowledge-producing sector (universities and R&D institutions) and the goods and services-producing sectors of an economy, such results might be expected.

Stable *vertical relationships* between users and producers can reduce the costs related to information and communication, the risks associated with the introduction of new products, and the time needed to move an innovation from the laboratory or design table to market. *Horizontal collaboration* between same-sector small and medium-sized enterprises can also yield 'collective efficiencies' (Schmitz, 1989) in the form of reduced transaction costs, accelerated innovation through more rapid problem-solving and greater market access. Agglomerations can generate *positive externalities*, such as the availability of skilled labor, of certain kinds of infrastructure, of innovation-generating informal exchanges and learning made possible through the adoption of conventions (Storper, 1995; Glasmeier and Fuellhart, 1996). These studies also stress the *supporting role that political and social institutions and policies play* in the development of partnering activity and in stimulating the transformation of such networks into broader systems of innovation and production at local, regional and national levels (Amin, 1999; Best, 1990; Piore and Sabel, 1984; Morgan and Sayer, 1988; Storper, 1993, 1997; Gertler, Wolfe and Garkut (1998).

Since the 1970s, governments in the industrialized world and more recently in a number of Third World countries, have come to believe that locational advantages such as those enjoyed by Japanese supplier networks or the firms in Silicon Valley or Northern Italy are critical for development. Governments at all levels – municipal, regional, national and quasi-supranational in the case of the European Union – began to directly encourage the formation of inter-firm collaborative agreements in R&D, production and marketing. The role of regional and municipal programs in promoting the development of technopoles in France, Spain and Portugal (Vavakova, 1988, 1995; Godinho,

Selada and Vedovello, 1997) national programs in promoting networking among small and medium-sized enterprises, and European programs such as ESPRIT and EUREKA (Mytelka, 1991) designed to stimulate research and technology development (RTD) and more recently to promote and support regional innovation (Morgan and Henderson in this volume) illustrate this movement.

Despite the rise in partnering activity and the development of new forms of supplier–client and user–producer relationships, there is remarkably little consensus on their contribution to innovation and hence to the growth of firms. Lined up on the positive side is the work of Ken-ichi Imai (1988a, b) showing how such networks developed strong regional poles in Japan, the arguments made by Sabel, Kern and Herrigel (1989) that the new supplier relations in the automobile industry are redefining the industrial corporation and as such are making a better contribution to development, and the detailed case studies by Kenney and Florida (1991) of the wide network of suppliers created by Japanese transplants in the automobile and steel industries in the American rust belt, all of which, however, are greenfield investments. Networking in the furniture industry, in textiles and clothing and in robotics all seem to have produced positive results for firms in Italy and Denmark (Belussi, 1987; Best, 1990; Camagni, 1986; Maskell and Malmberg, 1999) and a survey of 719 'cooperating' industrial companies in Northrhine Westphalia suggests that the growth of employment, turnover and exports has been higher in firms with R&D and marketing partnerships than in those without (Belzer, 1994). There is no indication in the study, however, as to whether these partnerships are local or long distance. Underlying the positive results in many of the most recent studies is the argument that tacit forms of knowledge are accessed and transferred more easily through face-to-face contacts not merely because proximity creates a propitious environment for such transfers but because they take place within the context of social networks based on trust and a shared culture (Maskell et al.,1998).[9]

On the other side are the skeptics – Dieter Ernst whose work on the electronics industry in Asia and Germany suggests that true partnerships are few and far between (Ernst and O'Connor, 1992; Ernst, 1997); Morgan and Sayer who argue that Japanese transplants in the electronics industry in South Wales did not develop the local supplier networks nor R&D linkages that were created elsewhere, in other industries and at other times, a point that comes up again in connection with the Asian experience with new forms of supplier–client relations in the textile and clothing and the electronics industries (Mytelka and Ernst, 1998); and lastly Edquist and Lundvall (1993) who, based on research in Denmark and Sweden, suggest that collaborative networks may only be useful for process-oriented, incremental change but not for breakthroughs.

Several factors of relevance to learning and innovation within regions and by small and medium-sized enterprises in particular, stand out in this literature.

First, there is some indication of a strong sectoral influence differentiating less knowledge-intensive industries such as food and furniture from information technologies. Second, some of the contradictory results obtained in these studies may be explainable in terms of a qualitative difference in the nature of the relationship depending upon whether it is a local or a long-distance partnership.

From a broader perspective, the ability of local economies to remain independent of the logic of multinational structuring of productive space in an era of globalization has also been questioned (Amin and Robins, 1990). The relationship between local and long-distance partnering would appear to be a key determinant here. So, too, however is the extent to which strategic partnerships are forming the basis for knowledge-based networked oligopolies on a global scale.

Knowledge-based networked oligopolies share four principal characteristics (Mytelka and Delapierre, 1999). First, they are knowledge-based, that is, they involve collaboration in the generation and use of, or control over, the evolution of new knowledge. As a result, the new knowledge-based oligopolies are dynamic, seeking to organize, manage, and monitor change as opposed to rigidifying the status quo.

Second, their focus is less on creating static size barriers to entry than on shaping the future boundaries of an industry and technological trajectories, standards, and rules of competition within it, which themselves are a source of dynamic entry barriers. In the Information and Communications Industries (ICT) of the 1990s these new rules included:

- innovation-based competition with rapid movement down the performance/cost curve;
- equally rapid movement down the manufacturing learning curve in order to ensure higher yields, rapid ramp up in volume to reduce costs;
- speed and flexibility in changing over to new product generations as the product life cycle shortens;
- increased use of M&As to extend product variety, assure brand-name recognition of products with the same basic functionality and gain market share in principal markets around the globe;
- increased use of strategic partnering to reduce the high costs and risks of R&D needed to maintain the pace of innovation, speed up the innovation process and shape the technological trajectory within an emerging industry or industry segment; and,
- efforts to maintain positions within the core group of firms in knowledge-based networked oligopolies through which the industry's future is increasingly shaped.

Third, knowledge-based networked oligopolies are composed of networks of firms rather than of individual companies. Alliances thus form the basic structure and building-blocks of the global oligopoly.

Lastly, in terms of their organization, the new oligopolies can form within or across industry segments and sometimes do both at the same time. They are including new actors whose assets complement the network and eliminating others whose resources are no longer critical (Mytelka and Delapierre, 1999).

Despite the powerful shaping effects of strategic partnering activity, M&As and the new knowledge-based networked oligopolies on a global scale, few studies of innovation systems – national or sectoral – have taken these interrelated phenomena into consideration. Worse still, FDI continues to be the object of bidding wars among countries and within them, across regions. The remainder of this paper explores these issues in greater detail.

Globalization, locational tournaments and local learning

Unlike the traditional industries that formed the core of Italy's most dynamic clusters, newer, more R&D-intensive industries such as those in the information technology industry, have less frequently replicated the Silicon Valley effect, constructed clusters such as Sophia Antipolis notwithstanding.[10] In R&D-intensive industries, it is possible that agglomeration effects at subnational levels may only be realizable if complemented by international networking.[11] The reverse, however, is not necessarily true, that is, international networking alone may not provide the requisite externalities to sustain the emergence of positive agglomeration effects when these may be needed for continuous innovation over the longer term. To some extent, both sets of considerations motivated the development of the European strategic program for research and development on information technology (ESPRIT) and the subsequent extension of such EU and broader European programs such as EUREKA to include telecommunications, biotechnology, environmental technologies, new materials and automotive technologies.

A comprehensive analysis of the substitutability of long-distance for local partnerships, particularly for small and medium-sized enterprises, is beyond the scope of this paper. However, an analysis of SME participation in long-distance partnering activity through the European RTD programs is suggestive, particularly when it is recognized that these partnerships significantly reduce the burden of financing both the joint research and technological development and the transaction costs involved in undertaking the collaboration. They also bring researchers into face-to-face contact over a number of years and thus provide a closer approximation of the kinds of trust-building relationships which are thought to stimulate learning and innovation.

Available evidence, however, shows that the level of participation by SMEs in such programs is relatively low. In the case of EUREKA Table 5.1 shows that SMEs accounted for between 26 and 32.5 per cent of the participants in any given year over the period 1989–93. Moreover, the 1995 evaluation survey of 452 participants in completed projects revealed that of the 319

Table 5.1   Participation by Industrial Firms in the EUREKA Program

Year	Number of projects	Number of large firms	Number of SMEs	SMEs as % of total firms
1989	292	708	319	31.1
1990	369	1,057	374	26.1
1991	489	1,206	460	26.6
1992	539	1,418	542	27.7
1993	75	1,576	757	32.5

Source: Commission: 1994, p. 232.

industrial respondents, 49 per cent were SMEs, which is well above their level of participation in all projects. But of these 156 small and medium-sized enterprises, 60 per cent were subsidiaries of large firms (EUREKA, 1995, p. 19). To put things in proper perspective, a survey of French SMEs showed that in 1993, 13 244 *independent* SMEs were involved in R&D (France, 1996, pp. 48–9).

Another way to measure the involvement of SMEs in intra-European networking is to look at the extent to which they are involved in multiple projects. In 1992–93, EUREKA did an evaluation of the Industrial and Economic Effects of its Projects. Table 5.2 provides data for the 935 respondents of which roughly 650 were industrial firms. What is remarkable about this Table is the evidence it provides of the limited ability of SMEs to sustain multiple long-distance partnership, though a large proportion of these firms engage in in-house R&D and other studies have shown the importance of networking and partnering activity for SMEs.

Data for major programs within the Second and Third Framework programs of the European Communities show similar results. Despite considerable effort to increase SME participation, the number of SMEs participating in shared-cost RTD programs such as ESPRIT, Race, Brite-Euram or Bridge, fell from 2,368 in the Second Framework to 2,174 in the Third Framework period (Commission, 1994, p. 231). Inclusion of the Craft program, set up to involve SMEs with limited R&D capabilities in cooperative research adds some 650 SME participants to the Third Framework program total (Commission, 1994, pp. 224–5). The level of SME participation in all programs under FPIII amounted to only 5,439 enterprises. Under the fourth framework program (FPIV), a number of new programs were added to support the participation of SMEs in EU RTD programs. These included the provision of legal advice, business planning and marketing advice, and partner search services. EU advisors also helped SMEs to draft all or part of their proposal, prepare consortium agreements, and access complementary information (Commission, 1999). This raised SME participation in cost-sharing programs in the first two years of FPIV (1994–96) to 2,801 enterprises with

Table 5.2 Incidence of Other Projects and Sources of Support (% of Respondents)

	EUREKA			EC			National			In-House		
	SMEs	Large	n.i.	SMEs	Large	n.i.	SMEs	Large	n.i.	SMEs	Large	n.i.
% in other projects	26	46	39	31	51	62	51	68	77	68	61	66

n.i. – research institutions, governments and higher education establishments
Source: EUREKA, 1993, p. 33.

overall participation of 5439 firms (Commission, 1997). Nonetheless, what these data suggest is that the number of independent SMEs involved in European programs is far too small to be a substitute for the dense fabric of local linkages which are believed to stimulate learning and innovation, particularly in the SME sector.[12] It is in this context that the impact of locational tournaments on local linkages becomes important.

Bidding wars add a number of new dimensions to the turbulence created by accelerated capital mobility and its impact on local linkages, especially for SMEs.[13] While competition among nation-states is not new[14] and incentives, in themselves, do not create the flow of foreign capital or wholly determine its location, three features of contemporary locational tournaments suggest that the nature of that competition has significantly changed. These changes might have a bearing on localized learning and innovation potential. They are: the widening range of incentives offered to non-local investors by an increasing number of *industrialized* countries based on static considerations of employment and exports;[15] the extension of competition to all levels of government;[16] and the diversion of attention away from the promotion of locally based activities and towards the massive use of local, state and national funds to attract investors of all stripes from elsewhere. A few examples of locational tournaments in Europe and the US will clarify the magnitude of the changes currently underway.

In the early to mid-1980s, Japanese automotive companies secured record subsidies in exchange for new plants in Ohio, Tennessee and Kentucky. State subsidies per employed person rose from $2,500 in the case of Honda's Marysville, Ohio plant in 1982, to $6,470 to Nissan (TN) in 1983, $14,263 to Mazda (MI) in 1987, $42,771 to Toyota (KY) in 1988, and $98,059 to Subaru-Isuzu (IN) in 1989 (Kenney and Florida, 1991, p. 30). Politicians at the time vowed to bring the bidding wars under control, but nothing came of it and over the 1980s the level of incentives offered to foreign firms escalated further.[17] By the early 1990s, as the number of cities and regions competing for each new plant increased,[18] firms were able to shift most of their up front costs onto the state thereby substantially reducing their risks (Mytelka, 1999).

Similar stories can be told about the electronics industry. Although the grant package might conceivably have been higher in Spain, north-east England, the runner-up in the contest for a Samsung electronics plant, was not without its benefits. Planned investment was £450 million and the subsidies included £58 million from the UK government and an additional £20 million from local authorities for a total of £78 million. This amounted to 17.3 per cent of total investment and £26,000 per job created (*Far Eastern Economic Review*: 12/22/94; UNCTAD, 1995, p. 295).

In the case of IBM's decision to locate its 64 Mbit Dram production at Corbeil-Essone in France, the investment was somewhat larger (about $1 billion), but the number of jobs it would directly generate was only 600.

This must be understood in the context of an earlier downsizing at this location that closed one semiconductor line in 1994 with a loss of 400 jobs and closures of production lines for bipolar and Cmos integrated circuits planned for 1996 and 1997 that would result in a further loss of 1,100 jobs.[19] The creation of 600 new jobs at this site, thus leaves a considerable shortfall. Nevertheless, the local (Corbeil), regional (Île-de-France), and national authorities in France competed actively to attract this investment away from rival sites in Germany (Dresden), Singapore and Taiwan. In total, financial incentives including reductions and exemptions from professional taxes and investment subsidies amounted to 300 million francs or 6 per cent of investment costs.[20] In addition, however, the national and regional governments have agreed to fund training activities and to subsidize the purchase of equipment by area research laboratories. According to one source, regional and national authorities have also extended the promise of contracts for the purchase of IBM PCs produced in France.

While locational tournaments and the subsidies they induce have traditionally been justified in terms of employment creation, the reality is that these jobs are being generated at high cost and with the possible exception of less favored regions where much of the investment is greenfield investment, in net terms, the increment is often quite small.[21] If from a short-term, allocative perspective one might seriously put locational tournaments into question, what can be said about the extent to which these newcomers generate 'growing territorial cores of learning-based industries . . . involv(ing) dense, local input–output linkages . . . (or) specialized knowledge and organizational talents . . . due to knowledge spillovers and complementarities . . .' (Storper, 1995, p. 13)? Data from the automobile industry provide both a temporal and contextual perspective in responding to this question.

Initially locational tournaments in the automobile industry could be justified on the grounds of employment creation. One of the most successful examples was the attraction of Nissan to northern England in the 1980s (Bridge, 1998). The range of incentives offered to BMW and Daimler-Benz in two recent cases were predicated on such earlier successes (Mytelka, 1999).

Recent changes in the nature of competition in the automobile industry during the late 1980s, however, have created contradictory tendencies with regard to the employment gains generated by bidding wars and put into question the contribution to local learning and innovation that foreign direct investments in the automobile industry had begun to show in the 1980s. Of critical importance in this respect is the emergence of preferred first tier suppliers with whom new forms of partnership were developed for the design of principal components and subsystems. The interactive nature of this relationship implied that preferred first tier suppliers were most often based in the home market of major automobile manufacturers. While in the past, most automobile assemblers had tended to create parallel supplier networks in each of the foreign locations in which they established major production facilities (Humphrey, 1998), the long-term contracts and special pricing arrangements with preferred first tier suppliers that emerged in the 1990s is inducing the latter to organize the supply of parts and subsystems in other locations (Le Gall, 1998; Young, 1996).[22] By reducing the number of suppliers and distinct components and parts, these partnerships have accelerated the pace at which new products are designed. Shared platforms, modularized production, long-term contracts with a global scope and the bringing of first tier suppliers within the assembler's own factory have further reduced costs and the uncertainties associated with a process of continuous change.[23]

This, in turn, has led to an acceleration in the pace of concentration within each of the industry's two main horizontal segments, auto parts and assemblers, and to the creation of new forms of partnership between them. M&As rose dramatically among automobile assemblers. This initially led to concentration within national markets and to the disappearance as independent producers of luxury car makers. More recently concentration has accelerated at the global level with the merger of Chrysler and Daimler-Benz, Ford and Volvo and Renault and Nissan.

Changes in competition and concentration within the automobile industry have contributed to the pursuit of aggressive market entry strategies on the part of the new auto parts system integrator firms to the globalization of this industry. Through mergers and acquisitions, auto parts manufacturers have also increased their size making it possible for them to take on a larger share of the design and manufacturing process and to extend the geographical scope of their activities.[24] Of the 620 automotive deals that were concluded in 1998, 320 involved parts suppliers.[25] Levels of concentration have increased within product categories and new horizontal segments are forming as 'system suppliers' to extend their production to cover whole subassemblies. In each of these modularized segments, consolidation is resulting in a relatively small number of top players. Car interiors were the first subassembly to be subcontracted. Today Lear Seating, Johnson Controls, and Forecia, each of which is the product of multiple M&As, along with captive suppliers, Delphi (GM) and Visteon (Ford) dominate this segment. In the engineering sector the market has similarly consolidated with Bosch, Denso, Dana, Magna and TRW as the principal independents alongside Delphi and Visteon in the manufacture of axles, steering and braking systems.

Under these new competitive conditions, bidding wars for automobile assemblers bring with them a host of secondary changes. These have the potential to disrupt existing ties between local auto parts suppliers and automobile assemblers. In so doing, they diminish the incentive to strengthen local R&D and design capacity to serve new investors. At the same time, however, the new relationship between assemblers and preferred first tier suppliers offers the possibility of raising the indirect employment gains from both expansion and new investment. It thus contributes to the intensification of bidding wars in this industry and to the contradictions they generate for local learning and innovation. Recent experiences in the UK, France and Brazil are illustrative of these contradictory effects.

In the late 1990s Toyota, then heavily invested in the United Kingdom, decided to introduce a new model. In competition with its Burnaston site in the UK, however, were sites in France, Germany, Spain, Austria and Poland. Penetration of continental markets was a key Toyota objective and Valenciennes, the chosen location in the Nord-Pas-de-Calais region, was on major transportation routes making supply and sales easier than for some of its rivals. But incentives were clearly a major factor in Toyota's decision to locate in France.[26] Located in a disadvantaged region, the Valenciennes plant was eligible for a wide variety of regional and national incentives. These reportedly include exemptions from some of the normal employers' social charges, reduced energy and land prices as well as grants from the French government amounting to nearly 20 per cent of the factory's building costs – a rate of subsidization double that which the United Kingdom's Department of Trade and Industry could offer. In exchange, the Toyota plant is expected to involve an investment of $655.6 million and create some 2000 direct jobs by the year 2005. Toyota has already invited its preferred first tier suppliers to locate in the region.

Unlike the North of England or Wales, Nord-Pas-de-Calais, however, was already a center for the production of automobiles and autoparts. Renault is the second largest employer in the region and the region is home to major autoparts subsidiaries of Renault and Peugeot such as ECIA (now Forecia), joint ventures of Renault/Peugeot (STA manufacturing gearboxes, Française de Mécanique motors), and many large and smaller French autoparts producers (Valeo, Plastic Omnium, Sotexo, a subsidiary of Bertrand Faure now Forecia). In 1996 and 1997 a number of American autoparts firms, Valmex (Textron), Japanese companies, Ogura, Akebono (Toyota Group), and US/Japanese joint ventures (Delphi (GM)/Calsonic (Nissan, now Renault)) have located in the region or announced plans to do so.

While these new investments will create jobs, unlike investments in 'greenfield regions' such as the earlier Nissan investment in the north of England, they may not increase overall employment in the region. Nor are they likely to strengthen the regional system of innovation. Most of the newcomers, as those above, will fall within the group of 'preferred' suppliers and their local joint venture partners. Their contribution to deepening learning and innovation within the region is likely to be small and a move in this direction by locally based firms has already been canceled.[27] The effect of having attracted a new Toyota plant to the region will also be felt by local suppliers indirectly as a result of the significant overcapacity in the European automobile industry, the age of existing Renault and Peugeot plants in the region and the difficulties which the latter face in securing fiscal and financial advantages similar to those which Toyota has obtained. This set of factors is likely to increase the vulnerability of local assemblers and their suppliers, reducing the ability of the former to maintain existing levels of production and sustain local linkages.

As in the EU, competition to attract foreign direct investment in the automobile industry was intense in other large markets around the world. The nature of changes in the automobile industry described above, moreover, heightened the interest of large automobile assemblers in penetrating these markets. Brazil was the largest of these. Over the 1990s, Brazil's provincial states, through the provision of massive incentives, succeeded in attracting a number of new automobile investors (NB: one, which had been attracted to the south has just renounced its investment in favor of moving to the North-East where incentives are even greater). What have been the consequences? Brazil, like France, had a well-developed automobile industry based on foreign assemblers but with a strong and growing domestic supplier industry. Large auto parts producers such as Metal Leve, Freios Varga and Cofap were known for their technological strength – all three were acquired by foreign companies in 1996 and 1997. A recent study of the impact of these takeovers has revealed that new investments are essentially import-intensive and are based on components designed and produced elsewhere. Technological activities in these companies has since been downgraded (Cassiolato, 1999). Even before these takeovers, assemblers had begun to shift component sourcing from local suppliers to first tier suppliers with production outside of Brazil. The import penetration coefficient for parts and components jumped from 8 per cent in 1993 to 20–25 per cent in 1996 (Cassiolato, 1999).

Conclusions

As the automobile case illustrates, inward foreign direct investment, at best, can complement and catalyze production locally and through its presence stimulate innovation through knowledge spillovers and the transfer of information and technology through supplier–client linkages. At its worst, it can crowd out local competitors, strip proprietary knowledge and other assets from these firms through mergers and acquisitions, and engage in a variety of market-distorting practices with highly negative effects for the achievement of broader social and economic goals. There is also increasing evidence that globalization of competition is pushing local SMEs to follow their clients overseas.[28] Global restructuring through both inward and outward FDI is thus leading to the rupture of networks that had been developed over the course of a decade or more.[29]

Changes in the mode of competition have accelerated the process of global mergers and acquisitions and have reinforced links between first tier suppliers and their clients in markets everywhere. As a consequence, the size barriers implicit in modularization and in the volume of purchases, the knowledge barriers resulting from the transfer of design to the component manufacturers, and the long-term and global nature of the contracts between auto parts manufacturers and automobile assemblers are becoming formidable barriers to entry for potential newcomers and for the survival of local independent suppliers throughout the world. This is particularly unfortunate as closer links between local parts suppliers and assemblers initially appeared to enhance the direct and indirect contribution of the automobile industry to employment and created the promise of building knowledge-based capacity and learning linkages within the local economy.

From a global allocative efficiency perspective, bidding wars are already problematic and evidence is accumulating that even from the perspective of job creation within given localities, the net gains are minimal. From a learning perspective, locational tournaments raise still more questions. For regional authorities, there is an obvious tension between the desire to strengthen local agglomerations and thus benefit from the positive externalities that these are said to generate and the desire to attract capital from elsewhere. This tension is particularly problematic in the context of what have been called learning economies, since these involve a process of learning and unlearning that takes place best through interaction. Developing and sustaining such interactions involves confidence and trust building and this requires both time and a multiplicity of opportunities for interaction. Accelerated capital mobility, exacerbated at the regional level by locational tournaments, it is suggested here, potentially erodes the basis for the development of such learning economies. Small and medium-sized enterprises, which are highly dependent on the localized presence of suppliers, clients, training institutions, R&D organizations and engineering firms, are thus particularly vulnerable to the turbulence created by locational tournaments. To the extent that long-distance partnering, such as that promoted by European RTD programs, does not substitute for the dense fabric of local linkages required by these enterprises, the problems SMEs face in competing in a globalized world are likely to become even more severe in the future.

Despite the ambiguous and often contradictory processes resulting from locational tournaments, most localities continue to actively attract investors with the offer of liberal incentives. There is still little awareness of the need to widen the view of benefits to be gained beyond employment and, to a lesser extent, exports. Long-term planning, which would take account of the sustainability of this employment in a world of global competition is not the strong suit in most regional and municipal investment promotion bodies. Nor, with rare exceptions, do regional authorities stress the need to strengthen local systems of innovation, though there is some evidence that regions which have succeeded in 'reinventing' themselves are those which have laid strong knowledge-bases and built a network of local linkages. As production becomes more knowledge-intensive and international capital flows continue their upward trend, sustained growth and development in local economies will require a rethinking of investment policies from a development perspective.

Policy-makers will thus be obliged to move beyond short-term considerations of employment and exports to a focus on the longer-term contribution that inward investment might make to learning and innovation. Attention will also have to be paid to policies and support structures that sustain the competitiveness of local SMEs even as they expand their activities abroad. A firm grasp on the changing nature of competition in globalized industries will be a fundamental prerequisite in designing effective policies that build a new, more positive, interface between the local and the global economy.

Notes

1 *See*, for example, Mytelka (1983, 1987); OECD (1992).

2 The term 'learning economy' is that of Bengt-Åke Lundvall (1995). *See also* Lundvall, 1988; Maskell and Malmberg, 1999; Nelson and Rosenberg, 1993; and Storper, 1995.

3 It has, however, attempted to do so. For example, learning and innovation have been dealt with in neoclassical terms by using rational choice models in the analysis of innovation (Lundvall, 1995, p. 6).

4 The term is borrowed from Paul David (1984).

5 *See*, for example, with regard to the advanced, industrialized countries, OTA, 1993; Reich, 1991; Stopford and Strange, 1991; Tyson, 1992; and the excellent summary of these positions in Ruigrok and van Tulder, 1994; and for Latin America, Devlin, Ffrench-Davis and Griffith-Jones, 1994.

6 There is no dearth of evidence for this proposition. With respect to restructuring in the UK textile and clothing industry, Diane Elson, for example, argues, that far more than a need to relocate labour-intensive segments of the textile industry to low-wage countries, pressure to delocalize in this industry during the 1980s was due to exchange rate fluctuations (Elson, 1989). In Korea, electronics exports received a new lease of life as a result of fluctuations in the dollar/yen rate during the early 1990s. By concealing fundamental weaknesses in the industry, the surge in exports delayed the implementation of key structural changes needed to sustain competitiveness in the future (Mytelka and Ernst, 1998).

7 *See also* Felix, 1995.

8 Cantwell and Sanna Randaccio (1992) argue that within the EU, oligopolistic rivalry under conditions of falling NTBs and for firms which do not invest strictly to serve the domestic market, that is, import-substituting M&As or greenfield investments, has led to considerable cross-investment and particularly cross-investment that is partially motivated by MNC interests in establishing EU-wide networks of technological activity (p. 104).

9 A study of the electronics cluster in Madrid, for example, concluded that a set of common values or a common ethic has not emerged among these companies and relationships between clients and their supplier networks tend to be hierarchical (Suarez-Villa and Rama, 1996, p. 1169).

10 Sophia Antipolis was created by the French government over 30 years ago and has been built around the transfer of R&D research institutions to the area and the localization there of a major IBM research facility. Despite having attracted numerous enterprises over the years, the dynamics of tacit knowledge transfer that have characterized Silicon Valley have yet to emerge in this and similar 'technopoles'.

11 There is some evidence for this in Storper, 1993.

12 Under FPIV a large number of innovative pilot projects at the regional level were undertaken under Article 10 of the European Regional Development Fund. For a discussion of such regional initiatives *see* Henderson and Morgan, 1999.

13 This and the following four paragraphs are adapted from Mytelka, 1999.

14 Interwar beggar-thy-neighbour policies were one of the factors leading to the postwar GATT liberalization process. Despite this process, competitive devaluations, the use of subsidies and the mercantilistically motivated imposition of countervailing duties have become commonplace (UNCTAD, 1994).

15 Whereas investment incentives were commonplace in developing countries during the 1960s and 1970s, what is notable in the 1980s and into the 1990s, were the large number of industrialized countries offering fiscal incentives. A recent UNCTAD survey of 103 countries showed that 22 of the 26 developed countries offered a reduction in standard income tax rate for foreign investors, 15, an accelerated rate of depreciation, 11 gave tax holidays and other sorts of fiscal incentives such as deductions from social security contributions and exemptions from import duties (UNCTAD, 1995, pp. 292–3).

16 This is leading to a number of aberrant results. St. Louis and Kansas City both in the State of Missouri, for example, were competing along with New York City to become the new headquarters for Trans World Airlines, then in bankruptcy. The two Missouri cities offered competing sets of incentives. St. Louis eventually won. As Kenneth Thomas points out (Thomas, 1994, pp. 2–3), 'From the point of view of Missouri unemployment, the same number of jobs would have come to the state', whether St. Louis or Kansas City had won. But each lost relative to a situation in which they had not offered additional incentives to those provided by the State of Missouri, because the cost of attracting these jobs rose. Moreover:

> (f)rom the standpoint of the United States as a whole, it is even more irrational. Not only would the same number of jobs have been generated in New York City, Kansas City or St. Louis, there are offsetting job losses at the company's former headquarters at Mt. Kisco, New York. Thus, sub-national governments prepared three sets of investment incentives to reward TWA for creating no new jobs in the U.S.

17 Governor Evan Bayh of Indiana, for example, won election in 1988 after accusing his predecessors of overpaying to attract Mitsubishi's Diamond Star plant in 1985, but by 1991, he was being attacked by neighboring governors for offering United Airlines $291 million in incentives to build a new maintenance facility in Indianapolis (*Wall Street Journal*, 25 November 1993). High levels of subsidization were also a feature of Hyundai's investment in Bromont, Québec (Rourke, 1989).

18 In the recent competition for a Mercedes-Benz plant some 170 cities and regions were initially in the running (*Wall Street Journal*, 25 November 1993). Although the runner-up to Alabama in the Mercedes case was another American state, North Carolina, the competition among municipalities and regions is increasingly international. Livingston, Scotland, for example, won out over Roseville, California in its bid to attract NEC Corporation's latest computer chip fabrication facility (*International Herald Tribune*, 22 September 1994).

19 These data were drawn from articles published in *L'Usine Nouvelle* (F) of 9 and 23 November 1995, *Le Figaro-Eco* (F), 4–5 November 1995 and *La Tribune* (F) of 3 November 1995 and 5 February 1996.

20   Less-favored regions must often bid higher to attract such investment. Thus Ireland offered Intel a subsidy equivalent to 18.5 per cent of investment costs for a plant that opened in Kildare in 1994 and a subsidy amounting to 11.8 per cent of investment for a second wafer fabrication plant in Kildare in 1995. This represents a subsidization rate per job created of close to US$80,000 (*Financial Times*, 19 Nov. 1995 and *Wall Street Journal*, 23 Nov. 1995).

21   Approximately 60 per cent of all FDI inward investment in Europe is ownership changing investment (UNCTAD, 1995) and much of it results in downsizing. IBM, in the course of its restructuring in Europe, sold off several of its plants and in each case there was a substantial reduction in employment. Nonetheless, even this investment is often subsidized with the subsidies justified as required to save jobs.

22   In the VW truck plant at Resende in Brazil, suppliers have been brought within the plant and actually carry out production using their own components and workers.

23   In the Information and Communications Technology (ICT) Industry, Hewlett Packard has begun to imitate this model.

24   Robert Bosch has bought a controlling interest in several South Korean firms. Mahle of Germany acquired Metal Leve of Brazil and thus gained access to both the large Brazilian automobile market and the design facilities of Metal Leve in the United States.

25   'Major Auto Mergers Drive Sweeping Change in the Parts Industry According to PricewaterhouseCoopers Survey', www.investing.lycos.com, 29 March 1999.

26   These data were drawn from articles published in the on-line *Daily Telegraph* of 18 March 1997, 8 May 1997, 21 October 1997, 9 December 1997, 10 December 1997, *The Financial Times* of 10 December, 1997 and the *International Herald Tribune* of 10–11 January 1998.

27   In 1996 Cofimeta, the automobile division of Arbel was planning to establish a center for machine design for the automobile industry in the Nord-Pas-de-Calais employing an expected 200 engineers and technicians. *L'Usine Nouvelle*, 'La France de l'Industrie' Edition, 1996, p. 144.

28   Six American parts suppliers, for example, purchased companies divested by Fiat and several others, component companies shed by Daimler-Benz. Reverse flows also exist. French autoparts suppliers, large firms such as Valeo and medium-sized companies such as Electricfil have invested recently in the United States, following American and European clients. (*L'Usine Nouvelle*, 22 février 1996, Les Echoes, 8 mars 1995).

29   The case of Montpellier in the wake of IBM's drastic reduction in operations is but one example (*L'Usine Nouvelle*, 31 août 1995).

References

Amin, A. (1999) 'An Institutionalist Perspective on Regional Economic Development', *International Journal of Urban and Regional Research*, vol. 23, no. 2 June, pp. 365–78.

Amin, A. and K. Robins (1990) 'The Reemergence of Regional Economies: the Mythical Geography of Flexible Accumulation', *Environment and Planning D: Society and Space*, vol. 8, pp. 7–34.

Amin, A. and N. Thrift (1994) 'Living in the Global', in Amin and Thrift (eds) *Globalization, Institutions, and Regional Development in Europe*, Oxford: Oxford University Press, pp. 1–22.

Belussi, F. (1987) 'Benetton: Information Technology in Production and Distribution: a Case Study of the Innovative Potential of Traditional Sectors', Brighton: University of Sussex, Science Policy Research Unit.

Belzer, V. (1994) 'Kooperationspraxis im Verbeitenden Gewerbe', Projektbericht IAT-DS 01, paper presented at the EUREKA Technology Conference, Lillehammer, Norway, June.

Best, M. (1990) *The New Competition Institutions of Industrial Restructuring*, Cambridge: Polity Press.

Bridge, J. (1998) 'Supply-Chain Dynamics – a Case Study of the Automotive Sector', paper presented at the International Workshop on Global Production and Local Jobs: New Perspectives on Enterprise Networks, Employment and Local Development Policy, Geneva, ILO, 9–10 March.

Camagni, R. (1986) 'Robotique industrielle et révitalisation du nord-ouest Italie', in J. Federwisch and H.G. Zoller (eds), *Technique Nouvelle, Ruptures Régionales*, Paris: Economica, pp. 59–80.

Cantwell, J. and F. Sanna Randaccio (1992) 'Intra-Industry Direct Investment in the European Community: Oligopolistic Rivalry and Technological Competition', in J. Cantwell (ed.), *Multinational Investment in Modern Europe*, UK: Edward Elgar, pp. 71–106.

Cassiolato, J. (1999) 'Asset Stripping FDI in Brazil', unpublished note, Rio, UFRJ.

Commission of the European Communities (1994) *The European Report on Science and Technology Indicators*, Luxembourg, DGXIII, October.

—— (1997) *Framework Program IV, SME Participation 1994–1996*, Brussels, DGXII, 18 February.

—— (1999) 'Small Companies Play a Bigger Part', *Innovation and Technology Transfer*, vol. 1 (January), pp. 11–12.

David, P. (1984) 'High Technology Centers and the Economics of Locational Tournaments', Stanford: Stanford University, mimeo.

Devlin, R., R. Ffrench-Davis and S. Griffith-Jones (1994) 'Surges in Capital Flows and Development: an Overview of Policy Issues', in R. Ffrench-Davis and S. Griffith-Jones (eds), *Coping with Capital Surges: Latin American Macroeconomics and Investment*, Boulder: Lynne Rienner.

Edquist, C. and B.-A. Lundvall (1993) 'Comparing Danish and Swedish Systems of Innovation', in R. Nelson (ed.), *National Innovation Systems: a Comparative Analysis*, New York: Oxford University Press, pp. 265–98.

Elson, D. (1989) 'Bound by One Thread: the Restructuring of UK Clothing and Textile Multinationals', in MacEwan and Tabb (eds), *Instability and Change in the World Economy*, New York: Monthly Review Press, pp. 187–204.

Ernst, D. (1997) 'Partners for the China Circle? The Asian Production Networks of Japanese Electronics Firms', in B. Naughton (ed.), *The China Circle*, Washington, DC: The Brookings Institution.

Ernst, D. and D. O'Connor (1992) *Competing in the Electronics Industry*, Paris: OECD.

EUREKA (1993) *Evaluation of EUREKA Industrial and Economic Effects*.

EUREKA (1995) *Eureka Evaluation Report*.

Felix, D. (1995) 'Financial Globalization versus Free Trade: the Case for the Tobin Tax', Geneva: UNCTAD Discussion Paper No. 108 (November).

France, Ministère de l'Education Nationale, de l'Enseignement Supérieur et de la Recherche (1996) *Recherche et Développement dans les Entreprises Résultats 1993*, Paris: Direction Générale de la Recherche et de la Technologie, janvier.

GATT (1993) *International Trade Statistics*, Geneva: General Agreement on Tariffs and Trade.

Gertler, M.S., D.A. Wolfe and D. Garkut (1998) 'The Dynamics of Regional Innovation in Ontario' in J. de la Mothe and G. Paquet (eds), *Local and Regional Systems of Innovation*, Amsterdam: Kluwer Academic Publishers.

Glasmeier, A. and K. Fuellhart (1996) 'What Do We Know about Firm Learning?' in European International Business Academy, Innovation and International Business, Stockholm: Institute of International Business, vol. 1, pp. 279–312.

Godinho, M.M., C. Selada and C. Vedovello (1997) 'Portuguese Technological Infrastructure: a System in Rapid Growth but in Need of Coherence', paper presented to the INTECH/EU Conference on Technology Policy and Less Developed Research and Development Systems in Europe, Seville, 17–18 October.

Henderson, D. and K. Morgan (1999) 'Regions as Laboratories: the Rise of Regional Experimentalism in Europe', Cardiff: Department of City and Regional Planning, Cardiff University.

Humphrey, J. (1998) 'Globalisation and Supply Chain Networks in the Auto Industry: Brazil and India', paper presented to the International Workshop on Global Production and Local Jobs: New Perspectives on Enterprises, Networks, Employment and Local Development Policy, ILY, Geneva 9–10 March.

Imai, Ken-ichi (1988a) 'Japanese Corporate Strategies toward International Networking and Product Development', paper presented to the Japanese Corporate Organisation and International Adjustment Conference, Canberra: Australian National University, September.

—— (1988b) 'Network Industrial Organization and Incremental Innovation in Japan', Tokyo: Hirtotsubashi University, Institute for Business Research, Discussion paper No. 122, May.

Kenney, M. and R. Florida (1991) 'Rebuilding the Rust Belt', in *Technology Review* (Feb./March), pp. 25–33.

Le Gall, S. (1998) 'L'industrie automobile: la décomposition de l'oligopole traditionnel', in Forum/Cerem (eds), *L'émergence d'oligopoles en reseau fondés sur la connaissance*, Paris: Commissariat Général du Plan, Mars.

Lundvall, B.-Å.(1988) 'Innovation as an Interactive Process: from user-producer interaction to the national system of innovation' in G. Dosi, C. Freeman, R. Nelson, G. Silverberg and L. Soete (eds), *Technical Change and Economic Theory*, UK: Pinter Publishers, pp. 349–69.

—— (1995) 'The Social Dimension of the Learning Economy', Denmark: Aalborg University, Danish Research Unit for Industrial Dynamics, Working Paper No. 96–1.

Maskell, P. and A. Malmberg (1999) 'Localised Learning and Industrial Competitiveness', *Cambridge Journal of Economics* 23, pp. 167–85.

Maskell, P. *et al.* (1998) *Competitiveness, Localised Learning and Regional Development: Specialisation and Prosperity in Small Open Economies*, New York and London: Routledge.

Morgan, K. and A. Sayer (1988) *Microcircuits of Capital: 'Sunrise' Industry and Uneven Development*, Boulder, CO: Westview Press.

Mytelka, L.K. (1983) 'Le capitalisme fondé sur la connaissance et le changement dans les stratégies des entreprises industrielles', *Etudes Internationales*, vol. XIV (3), (Sept.) pp. 433–52.

—— (1987) 'The Evolution of Knowledge Production Strategies within Multinational Firms' in J. Caporaso (ed.) *A Changing International Division of Labour*, Boulder, CO: Lynne Rienner, pp. 43–70.

—— (1991) 'States, Strategic Alliances and International Oligopolies: the European ESPRIT Program' in L.K. Mytelka (ed.), *Strategic Partnerships and the World Economy*, London: Pinter Publishers, pp. 182–210.

—— (1999) 'Locational Tournaments for FDI: Inward Investment into Europe in a Global World' in N. Hood and S. Young (eds), *The Globalization of Multinational Enterprise Activity and Economic Development*, UK: Macmillan.

—— and M. Delapierre (1999) 'Strategic Partnerships, Knowledge-Based Networked Oligopolies and the State' in C. Cutler, V. Haufler and T. Porter (eds), *Private Authority and International Affairs*, Binghamton: SUNY University Press, pp. 129–49.

—— and D. Ernst (1998) 'Catching Up, Keeping Up and Getting Ahead: the Korean Model under Pressure', in D. Ernst, T. Ganiatsos and L. Mytelka (eds), *Technological Capabilities and Export Success in Asia*, UK: Routledge.

Nelson, R.R. and N. Rosenberg (1993) 'Technical Innovation and National Systems' in R. Nelson (ed.) *National Innovation Systems: a Comparative Analysis*, Oxford: Oxford University Press, pp. 3–22.

OECD (1992) *Technology and the Economy – the Key Relationship*, Paris: OECD.

Office of Technology Assessment (OTA), US Congress (1993), *Multinationals and the National Interest, Playing by Different Rules*, Washington, DC: Government Printing Office.

Piore, M.J. and C.F. Sabel (1984) *The Second Industrial Divide: Possibilities for Prosperity*, New York: Basic Books.

Reich, R.B. (1991) *The Work of Nations*, New York: Vintage.

Rourke, P. (1989) 'Hyundai and the Canadian Automotive Industry: the Influence of Canadian State Policies', M.A. Research Essay, Ottawa: Carleton University, Norman Paterson School of International Affairs.

Ruigrok, W. and R. van Tulder (1994) *The Logic of International Restructuring*, London: Routledge.

Sabel, C., H. Kern and G. Herrigel (1989) 'Collaborative Manufacturing: New Supplier Relations in the Automobile Industry and the Redefinition of Industrial Co-operation', Cambridge, MA: MIT paper.

Schmitz, H. (1989) 'Flexible Specialization. A New Paradigm of Small-scale Industrialization', Discussion paper 26.1, Brighton: Institute of Development Studies, University of Sussex.

Stopford, J. and S. Strange (1991) *Rival States, Rival Firms: Competition for World Market Shares*, Cambridge: Cambridge University Press.

Storper, M. (1993) 'Territorial Development in the Global Learning Economy: the Challenge to Developing Countries', unpublished paper.

—— (1995) 'Regional Economies as Relational Assets', paper prepared for presentation to Association des Sciences Régionales de Langue Française, Toulouse, 30 August–1 September.

—— (1997) *The Regional World*, London: The Guilford Press.

Suarez-Villa, L. and R. Rama (1996) 'Outsourcing, R&D and the Pattern of Intra-metropolitan Location: the Electronics Industries in Madrid, *Urban Studies*, vol. 33 (7), pp. 1155–97.

Thomas, K.P. (1994) 'European Union Regulation of Competition for Investment: Lessons for North America', paper presented to the International Conference on Economic Integration and Public Policy: NAFTA, the EU and Beyond, Toronto, 27–29 May.

Tyson, L. (1992) *Who's Bashing Whom*, Washington, DC: Institute for International Economics.

UNCTAD (1993) *Trade and Development Report, 1993*, New York: United Nations.

UNCTAD (1994) *Trade and Development Report, 1994*, New York.: United Nations.

—— (1995) *World Investment Report Transnational Corporations and Competitiveness*, NY and Geneva: United Nations.

—— (1998a) *Trade and Development Report, 1998*, New York: United Nations.

—— (1998b) *World Investment Report*, New York and Geneva: United Nations.

—— (1999) *World Investment Report 1999*, New York and Geneva: United Nations.

UNDP (1994) *Human Development Report 1994*, New York: Oxford University Press.

Vavakova, B. (1988) 'Technopole: des exigences techno-industrielles aux orientations culturelles', *Culture Technique*, Revue de l'Ecole des Mines, No. 18.

—— (1995) 'Building "Research-Industry" Partnerships through European R&D Programs', *International Journal of Technology Management*, vol. 10 (4/5/6), pp. 567–86.

Young, L. (1996) 'Auto Companies Bring More Suppliers in Early', *Electronic Business Today*, June, pp. 61–4.

Young, S., N. Hood and E. Peters (1994) 'Multinational Enterprises and Regional Economic Development', *Regional Studies*, vol. 28 (7), pp. 657–77.

6
Technology, Culture and Social Learning: Regional and National Institutions of Governance[1]

Meric S. Gertler

A new paradigm for regional development?

The last fifteen years of the twentieth century have seen the emergence of tremendous excitement, in both academic and policy realms, about the economic role of the region in advanced capitalist economies (Sabel, 1989; Cooke, 1999; Saxenian, 1994; Storper, 1995; Scott, 1996; Scott, 2000). Over this period, much has been written about the technological dynamism and innovation-fostering properties of regionally anchored economic systems. At a time when the terms of capitalist competition are said to have undergone a fundamental shift in favour of quality, innovativeness, responsiveness to market trends, timeliness, and the growing importance of intangible assets (Best, 1990; Leadbeater, 1999), the most successful of these regional clusters have attracted considerable attention due to their apparent ability to nurture groups of firms capable of generating new and innovative products in a timely fashion. Furthermore, what has most captured the imagination of scholars is the highly social nature of economic activity in these regions (Sayer and Walker, 1992). Not only have production systems become highly vertically disintegrated and transaction-intensive, but the non-market forms of interaction between firms – what Dosi (1988) and Storper (1995) have referred to as untraded interdependencies – are now seen to be as important as (if not more important than) actual market exchange. Perhaps the most important of these forms of social interaction is the process of social learning which arises between individual economic actors in such regions.

Holding these regional production and learning systems together are sets of institutions – public and private, formal and informal – which facilitate the close interaction and cooperation necessary to support innovation-based, knowledge-intensive production. Within such networks of firms, regionally based institutions are said to be responsible for creating and maintaining robust and effective mechanisms for the governance of behavior of

111

economic actors (Cooke and Morgan, 1998). Moreover, such governance mechanisms are said to foster openness and the relatively free exchange of proprietary technical information that is necessary to undergird a socially organized system of learning and innovative production through the generation of trust and social capital (Sabel, 1992; Putnam, 1993). A region's system of governance encourages the widespread adoption of and adherence to a set of shared norms, practices, attitudes and expectations amongst the area's individual firms and managers which lead to the formation of a collective order. This amounts to a set of rules shaping interfirm interaction and, in particular, fostering a socially based technology development system characterized by rampant, mutual interfirm learning. In such systems, the transfer of technology between individual firms is facilitated by both the relatively open, trust-based relations between firms and the fact that such regions do themselves evolve into leading sites of new technology production.

In the last few years, scholars studying this phenomenon have drawn considerable inspiration from the earlier literature on national innovation systems as developed by Freeman, Nelson, Lundvall and others (*see* Nelson, 1993). Given the recent interest in subnational institutions supporting the collective order of knowledge-intensive production systems, these researchers have developed the concept of regional innovation systems to capture the collection of firms, institutions and rules produced at the scale of the individual region, which shape the innovative behavior of the region's firms, managers and workers (Braczyk *et al.*, 1998).

Following such stories of impressive success in the international literature on economic development, it should come as no surprise that policy-makers have sought to emulate these achievements in less fortunate regions (Bosworth and Rosenfeld, 1993; Coopers and Lybrand, 1994; Staber *et al.*, 1996). In the process, a whole new paradigm of industrial development has emerged. By creating the right mix of regional institutions, or simply by exhorting firms to cooperate more (perhaps by demonstrating to them the benefits of collective action), regional development agencies have adopted the role of catalysts or *animateurs* (Morgan, 1996). Their objective in assuming this new role is to alter the behavior of firms, managers and workers in their region so that cooperation becomes as commonplace as competition between firms. With respect to innovation and technology transfer, a specific goal is to transform local firms into more innovative enterprises by 'opening them up' to a socially organized system of learning-based innovation (Lundvall and Johnson, 1994). A second goal is to raise the general standard of technical sophistication of local (usually small and medium-sized) enterprises by stimulating the dissemination of information about 'best-practice' production methods, either through the creation of demonstration centers and industrial modernization 'extension' services (Shapira, 1990; 1996), or by attracting technologically progressive lead firms to the region from afar and encouraging them to create opportunities for active learning between themselves and local supplier firms (Morgan, 1997).

In this chapter, I wish to argue that some of the premises underlying this emerging paradigm of industrial development require rethinking. First, I shall argue that the common understanding of the 'technology transfer problem' inherent in this approach is incorrect. Second, I shall argue that, while the concept of regional innovation systems is of fundamental importance, policy interventions which address the 'problem' of industrial innovation only at the level of the firm or the region will meet with limited success at best. This is because of the continuing importance of certain systemic influences defined by the national system of economic regulation (including, but not limited to, the national system of innovation) which shape the background conditions within which social learning processes might unfold.

In the following section, I shall examine the prevailing conception of the technology transfer problem in the context of regional development, offering a critical assessment of the now dominant view. In the third section of this chapter, I briefly review some of the findings of a multi-year study of technology transfer between German manufacturers of advanced industrial machinery and their Canadian customers (manufacturers of a range of other industrial and consumer products). The findings from this study serve to illuminate the mix of forces shaping (and greatly complicating) the social processes of innovation, including the important role played by apparent 'cultural' differences between technology producers and users. On the basis of this study, I hope to demonstrate that transferring technology (in this case, advanced process technologies) and collaborative work practices based on social learning is far more difficult than is generally assumed, for reasons that are not widely appreciated or well understood. We shall also come to appreciate the continuing importance of national regulatory features – mechanisms of governance – which influence the behavior of users and producers, reproducing and accentuating their 'cultural' differences over time, with major implications for the ease with which successful social learning dynamics might actually arise. In the final section of this chapter, I consider the more important implications for regional development and technology policy flowing from the above arguments.

Technology transfer, user–producer interaction, learning and tacit knowledge

The recent history of technology adoption experiences amongst firms in mature industrial regions of the United States, Canada and Great Britain reveals this process to be far more difficult, costly, and fraught with disappointment than was initially anticipated. When one reviews the experiences of new technology users (and would-be users) across a range of different regions and countries (for a survey of the international experience, *see*

Gertler, 1993), it is evident that the greatest difficulties seem to have been encountered by manufacturers in older, more mature industrial regions still dominated by earlier technological paradigms, or by manufacturers in small centers and rural locations. In such cases, the 'implementation problem' has most commonly been perceived by industrial policy-makers as one of stimulating and assisting the diffusion of new process technologies through manufacturing 'extension' and technology transfer programs (Shapira, 1990; 1996).

A growing number of state/provincial and national jurisdictions in the United States, Canada, and other countries launched modernization programs during the late 1980s and 1990s to help manufacturers upgrade their operations. Such programs have most commonly focused on assisting firms with the implementation of advanced machinery and production systems. However, in many cases these programs have been enlarged to include two other distinct but interrelated dimensions of change: (i) the adoption of a variety of forms of 'workplace reorganization' (for example, teams, simultaneous engineering, total quality management), and (ii) the promotion of network relations (that is, cooperation, knowledge-sharing and longer-term relationships) between client firms and their customers, suppliers and perhaps even competitors. If one regards 'technology' as a social phenomenon rather than purely the domain of the engineer, then all three of these dimensions can be thought of as part of the technology transfer process.

Such programs are typically delivered by using one of two principal methods. In the first case, technology transfer centers are set up for the purpose of allowing potential user firms to *learn about* the latest process (and often, product) technologies, and to receive assistance in *learning how to use* such technologies properly. Training is frequently provided for a client firm's machinery operators. Furthermore, such demonstration centers are usually located within close proximity to a group of target client firms in a particular region. Often, the programs offered by such centers will be specially tailored to meet the needs of one or more industrial sectors (for example, metalworking, woodworking, plastic products) prevalent in the region.

A second common approach, modeled on the decades-old concept of 'agricultural extension', uses mobile consultants who travel to the user firm's plant. These 'manufacturing extension' agents are usually trained in engineering or other relevant technical and scientific disciplines, and give specific advice and assistance to help solve technology implementation problems. This approach underlies the program greatly expanded by the Clinton Administration in the United States. Known as the Manufacturing Extension Partnership, this program was set up to improve the competitiveness of SMEs by providing funding to state governments to help local SMEs adopt 'best-practice' production methods (Gittell *et al.*, 1996; Shapira *et al.*, 1996). Where modernization programs also aspire to developing network-style learning relations between firms in a particular region, the extension approach is employed to help link up individual firms with potential partners for cooperation. Those agents responsible for engineering such match-ups are often referred to as 'network brokers', for obvious reasons (Bosworth and Rosenfeld, 1993).

With the benefit of hindsight and several years' experience with such programs, the limited success of this form of intervention suggests that it has been driven by a poorly specified model of the process by which technologies are developed and implemented. One alternative way of conceiving of this process is to view it from the perspective of interaction between technology users and technology producers (Lundvall, 1988; Gertler, 1993; 1995). From this user–producer interaction perspective, such implementation difficulties arise because of insufficiently close interaction or ineffective communication between technology users and the firms producing these technologies. Moreover, failure to achieve such close interaction can be anticipated in those cases where user firms are located in regions not heavily endowed with producers of process technologies.

It is worth examining these arguments closely, since their insights bring us considerably closer to a proper understanding of the technology transfer process and the reasons why difficulties arise under particular circumstances. The central idea is that complex production technologies are not only more likely to be *adopted* successfully when there is close and frequent interaction between producer and user, but are also likely to be *produced* more successfully as well. This interactive mode of technology acquisition allows users to gather as much knowledge as possible about the properties of the machinery under consideration, and to gauge the reliability and trustworthiness of the producer. Furthermore, it may allow the user to make its technological needs more readily and clearly known to the producer, creating the conditions under which the effective customization of the technology to the user's particular application is more likely. To allow customization to occur, however, users must reveal to an outside firm certain proprietary details concerning their products or production processes, and they may be unwilling to do so unless they have been able to build up a sufficient level of trust with machine producers, resulting from a process of close interaction over an extended period of time.

For these reasons, this literature attributes considerable importance to cultural commonality or proximity, owing to the fact that much of the technological capability and know-how embodied within advanced machinery is produced through tacit rather than explicit means (with tacit knowledge pertaining to the user's needs being transmitted from the user to the producer to enable the development of a more effective design). Furthermore, the proper use of such machinery may also depend on the effective transmission of tacit knowledge from producer to user. Hence, Lundvall (1988) gives considerable significance to culture, arguing that a common cultural background and language facilitates the kind of detailed, effective communication that is crucial to the success of user–producer interaction. While this implies that the nation-state is a key geographical unit in understanding this relationship, even closer contact might be required (p. 355):

> In the absence of generally accepted standards and codes able to transmit information, face-to-face contact and a common cultural background might become of decisive importance for the information exchange.

To put this in more contemporary terms, the kind of interaction described above is a classic example of innovation based on interactive learning or, to use Lundvall and Johnson's (1994) expressive phrase, 'learning-by-interacting' (*see also* Morgan, 1996). In recent years, the concept of tacit knowledge and its role in the technology development and transfer process has received a considerable amount of well-justified attention (Storper, 1992; Nonaka and Takeuchi, 1995). What are the implications of this work for our understanding of the technology transfer process? The most direct implication would seem to be that the technology transfer and implementation difficulties discussed earlier can now be understood as arising from insufficient opportunities for technology users and producers to engage in interactive learning and the effective sharing of tacit knowledge. If true, then the obvious implication (either for firms with sufficient resources or for development agencies intervening to assist firms in difficulty) is to resolve such problems by pursuing more extended, frequent face-to-face contact between technology users and producers. This might be achieved either through location in the same region (an approach which has inspired governments to stimulate the development of an indigenous machinery-producing sector), or through the support of frequent, sometimes lengthy visits by producers to their users' plants (or vice versa). If the latter option was pursued, perhaps by giving 'interaction grants' to SMEs, to accompany other direct subsidies to capital investment, then this would work best if it occurred during all three phases of the technology production/transfer process, recognizing its extended, interactive, non-linear nature.

These insights take us several important steps closer to a sound understanding of the nature of technology transfer and the circumstances under which it is most successful. However, they still stop short of a complete answer. In particular, the 'tacit knowledge problem' only goes some way toward capturing the foundations of effective technological interaction. I shall argue in the remainder of this chapter that it is not simply the opportunity for sufficiently close contact between firms (afforded by physical proximity) sustained over time which underlies this process. Nor is it strictly a matter of achieving cultural or linguistic commonality in order to facilitate the communication and sharing of technological subtleties. In the canonical cases of successful, regionally based technology transfer networks – such as those in Germany's Baden-Württemberg – physical proximity between users and producers, a shared cultural heritage, and linguistic commonality undoubtedly exist and are important to the internal workings of such districts. However, there is much more than this that facilitates the cooperative interaction for which such regions have become famous. I shall argue below, using a transnational case study for illustrative purposes, that these more obvious forms of commonality mask another, more fundamental influence of key importance to the success of such social learning in regional economies: namely, a shared set of rules, expectations, and norms which arise not simply from a common history and culture but from a common macro-regulatory framework. Furthermore, this framework (in contrast to others) is one which is *conducive to cooperation between firms*. It is also shaped largely through the institutional forms and interventions of the nation-state.

Cross-national technology transfer and learning: geographical, cultural or institutional divide?

Probably the most distinctive feature of Canada's industrialization during the twentieth century was the reliance of its manufacturers and resource companies on foreign sources of production technologies (Britton and Gilmour, 1978; Porter, 1991). This legacy is still strongly evident today, as the dominant sources of those advanced manufacturing technologies in use within Canadian manufacturing plants are found outside the country: in Japan, Germany, Italy, Sweden and other European and Asian countries. For over a decade, I have been studying the process by which manufacturers in Ontario have acquired and implemented these new process technologies. This program of research has focused on the relationship between these user firms and the companies which produce their production technologies. I have conducted an extensive survey of users in Ontario, follow-up interviews with a representative subset of these users, and extensive interviews with their technology producers in one of the principal supplier countries: Germany (*see* Gertler, 1995; 1996 for further details). In addition, I interviewed a number of representatives and suppliers of these foreign machinery producers, residing in Canada and serving as intermediaries between overseas producers and local users.

The findings indicate that many users in Ontario continue to experience significant problems of implementation and operation, long after the installation is completed (for a more complete discussion, *see* Gertler, 1995). Even after being given time to 'move along the learning curve', many users – including some large, relatively sophisticated operations with deep financial resources and plentiful in-house technical staff – have had considerable difficulty achieving effective implementation. The machinery and systems, once installed, frequently failed to live up to the user's expectations (or the salesperson's claims) for product flexibility, speed of production and changeover, quality, ease of use and reliability. Breakdowns and malfunctions were frequent and downtimes were lengthy and disruptive. In general, the returns from such costly and difficult investments were usually disappointing. Further-more, and crucially (given the theme of this chapter), these problems seem to have been particularly likely to arise (and to be especially acute) when the technology in question originated in 'far-off' places such as Germany, Japan, and other overseas sources.

When asked to explain the reasons for and sources of these difficulties, both users *and* producers commented on the complications introduced when trying to carry out communications and transactions involving com-plex technical subjects over long distances – referring to both the initial spec-ification of technology requirements and the subsequent problem-solving and 'trouble-shooting' procedures. These complications were seen as stemming from the delays introduced by intervening time zones, the difficulties of technical problem-solving without face-to-face contact (despite the wide-spread use of information and telecommunication technologies to connect users to producers), and problems of comprehension which inevitably arise due to differences of language. These concerns address many of the issues raised by Lundvall (1988) and others working in the user–producer interac-tion paradigm.

However, further discussion revealed that a deeper source of difficulty lay in the fundamental differences in expectations, characteristic workplace practices and norms, attitude, managerial routines and transactional behav-ior – in short, what appear to be substantially different industrial or business 'cultures' in Canada and Germany. Indeed, interviewees on both sides of the Atlantic readily identified differences in culture or 'mentality' as the root of their problems in dealing with one another. Nevertheless, notwithstanding this widely shared diagnosis, more detailed analysis of specific instances in which these differences have become evident reveals that underlying these apparently cultural gulfs are fundamentally different regulatory regimes and institutional structures which are instrumental in producing and reproduc-ing these 'cultural' differences. Specific manifestations of this are presented below, expressed as differences in expectations, attitudes, accepted business customs and practices which have led to misunderstandings, disappoint-ments, conflict and, in extreme cases, termination of the relationship between technology producer and user.

1. Purchasing Practices

The process of purchasing advanced machinery, as suggested in our earlier review of the literature, normally takes place over an extended period. Once a user has identified a preferred supplier/producer, discussions between the transacting parties address the full range of details concerning the techno-logical features and performance characteristics of the machinery. At this stage, design specifications are worked out through an iterative process of interaction and discussion (itself, an important form of social learning). Commonly, the prospective user will also visit the plant of another user who has already implemented the same or similar machinery, in order to learn more about its operational characteristics, performance and reliability. However, there are significant differences in the manner in which customer/users from Germany and North America approach this process. In the case of Canadian and American users, typically only production managers and engineering staff participate in this process. Machinery vendors do not interact directly with shopfloor employees until well after the sale is made and technical specifications have been set. This contrasts sharply with the prevailing practice amongst German user firms, where production managers and shopfloor workers routinely participate *jointly* in the process of vendor selection and machinery design or customization. As a consequence, since the ultimate operators are involved in the design process, they are able to contribute their considerable detailed knowledge of production processes (both explicit and tacit) to the discussion, enriching the interaction and improving the performance characteristics of the resulting machinery. These workers will also tend to accept the introduction of such new process technologies with enthusiasm, since they played an integral process in their initial selection and design.

2. *Contracting Procedures*

German machinery producers reported that North Americans 'do business differently' from the way it is normally done in Germany, and even in other parts of Western Europe. At issue here is the degree of formality of the agreements surrounding a commercial transaction. Generally speaking, North American customers were seen to insist – as a rule – on a formal, legal contract to bind a sale. This came as a surprise to many German firms when they first began to deal with North American customers in significant numbers. Furthermore, such contracts, while not completely unheard of in Europe, were nevertheless considerably more extensive and inclusive than those which they had signed with European customers. They would often include not only the standard payment schedules (usually incorporating staged payments), but also contained clauses about vendor-provided training and warranty length and conditions. Furthermore, they incorporated performance clauses in which final payment would be contingent upon the customer's satisfaction with operation once the system had been installed in the user's plant. This insistence, while commonplace in North American contracts, often offended the German producers who prided themselves on the technical excellence of their machinery, on their ability to solve any operational problems, and on the strength of their verbal commitments to stand by their machines (and their customers).

3. *Payback Periods*

Canadian and American customers expected and demanded that new capital goods would pay for themselves within one to three years, reflecting the heightened time pressures and shortened product life cycles they are facing.

The German machinery producers generally found this a strange and surprising requirement (though they were usually unwilling to admit this to their customers), as their domestic users would rarely if ever make such demands. Nevertheless, they would often give assurances to their North American users that such payback schedules could be met, based on their experience with the performance of similar installations they had completed in Germany and the rest of Europe. However, this frequently proved to be incorrect, as the subsequent experience would show, and could later lead to discord. Indeed, one German manufacturer of advanced metal-cutting machine tools indicated that they got themselves into several serious binds over precisely this issue (as well as related promises over ease of operation and rates of output) in their first dealings with American customers ('Our sales people there made promises which can't be kept'). The firm's response was to stop selling their most advanced machines to American customers, since this technology could not be adequately implemented and supported from the firm's German base. Instead, they developed a simpler, 'stripped-down' version of their machinery expressly for the US market (*see* Gertler, 1996).

4. Training and Labor Practices

Here, too, the typical Canadian or American customer had very high expectations about vendor-provided training before, during, and after installation. They expected training to be lengthy, extensive, and starting almost from first principles of machinery operation. They also expected all of this to be provided as part of the previously agreed price for the machinery. On the other hand, the German producers had assumed that the training they were to provide would be complementary to (not a substitute for) user-provided training or prior worker knowledge acquired through formal instruction and on-the-job learning. Hence, when producers provided training, they tended to pitch their instruction at a level that was too advanced for most of the customer's workers and managers, based on the expectations they themselves had formed through their experience with their German and European customers' workforces. Furthermore, when this difference in starting levels of competence became apparent, and the producer realized just how much 'remedial' instruction would be necessary to bring the customer's workers up to speed, they would become concerned to recover some of the costs associated with the additional training required. This too would lead to conflict more often than not.

Looking at labor practices more broadly, the rate and frequency of turnover in the labor force was much higher in North American customers' plants than it was in the plants of German users. Moreover, this was true for both shop-floor workers as well as management personnel. This had at least three effects. First, it meant that there was a far weaker incentive for North American customers to invest in training their own workers, since these employees might not remain with the company for very long, and could very likely end up working for a competitor. Second, it meant that the vendor (producer) was called upon to take on the major responsibility for training (see above), but would be frustrated in its efforts to do so, since the workers they had trained would not remain with the customer's firm for long. Indeed, once these workers had been trained, their value to other potential employers in the same region would have increased substantially, making them especially attractive candidates for poaching. Hence the necessary knowledge for successful operation of the machinery, which may have accumulated with the original vendor-trained worker, was now lost to the firm. Replacement operators rarely had sufficient knowledge to run the machinery properly, leading inevitably to breakdowns. It was frequently necessary for customers to incur further training costs to recapture this ability, although they tended to view this as the producer's responsibility (even long after vendor warranties had expired). Third, because the average length of job tenure was considerably shorter in the North American users' plants, this meant that customers' workers were afforded less opportunity to move along learning curves and acquire full knowledge about the effective use of the machine through on-the-job experience and learning by doing. This further escalated training needs, and compounded implementation difficulties.

5. Operating Manuals

German producers were also somewhat surprised that their Canadian and American customers would expect extensive, detailed and explicit printed manuals to guide them in the normal operation and debugging of their production systems. The machinery producers had rarely been called upon to provide this kind of information to their home customers and, as is evident from the preceding discussion of training, did not really know what it was that the user needed to know. They would make assumptions or guesses about the prior level of comprehension amongst management and workers in the user's plant which usually turned out to be inaccurate.

North American users were described by German producers as having a different 'mentality' or 'understanding' of technology itself. Moreover, this difference was usually not fully appreciated by the producer until the machine in question had already been designed and built. These customers would generally expect the full capabilities of the technology to be codified, 'locked into', and wholly *embodied* within the machinery and software itself. Whatever information could not be embodied in this way was expected to be provided on a continuing basis by the producer, either in the manual, or through regular and frequent service calls (see below). This attitude struck German producers as distinctly foreign and different from their own (and that of their European customers), since prevailing practices in domestic plants emphasized a more inclusive, *social* model of technology implementation, in which workers and managers cooperated to reap the full technological potential of their production systems.

6. Service Expectations

German producers reported utter surprise at the level of service expected and demanded by North American customers – that is, rapid response on a 24 hour, 7 day a week basis. Most producers regarded this level of expectation as an order of magnitude greater than the prevailing norm in Europe. As noted earlier, this need arises in part from high rates of labour turnover, the 'embodied' approach to technology implementation characteristic of North American users and, as we shall see below, their tendency to discount the importance of regular maintenance. Needless to say, most German producers were (at least initially) quite unprepared to provide this level and intensity of support and assistance, especially on an overseas basis. Not surprisingly, this requirement led many such firms to retain local service representatives in Canada or the US, or to establish parts and service subsidiaries there.

7. Maintenance Practices

One of the clearest differences to emerge was over the contrasting practices of German and North American users regarding machinery and equipment maintenance. German producers remarked (usually with disbelief and more than a little disdain) that North American industrial culture did not seem to assign much value to the importance of regular, preventive maintenance. As a consequence, production systems in Canadian and American plants would, in the view of the producers, fail with predictably greater frequency. This stood in sharp distinction to the dominant practice in German plants, where not only managers but also the operators themselves would maintain and service the machinery on a regular basis. More than one German producer commented on how, in their German customers' plants, the operators were 'married to' or 'owned' their machines, and would lavish attention upon them. In the words of one German manager, 'German workers . . . have the feeling "that is my machine, and I am responsible for it"'.

8. Operational Success

Canadian users complained that the German production systems were considerably more difficult to operate effectively than they had been led to believe at the time of sale. Even in cases where the user had travelled to another user's plant (often in Germany) to observe the real-time operation of a similar system before deciding to make the purchase, these post-implementation difficulties still arose. A frequently heard comment (both from those users that did buy German machinery and those that did not) was that German technology was 'too complex' and 'overengineered'. More often than not, users attributed this to deeply engrained cultural traits which predisposed German producers and users to overly complicated technical solutions – a kind of technology fetish. The German producers had difficulty knowing how to regard such complaints, since they were aware that similar systems worked perfectly well and with little difficulty in the plants of their German and other European customers. Instead, in the face of criticism from users that producers were 'rigid', 'unbending', or trying to 'dictate' inappropriate technical solutions to their precise production problems, the German firms would tend to place the blame with the user, accusing them of not doing enough training of its workers and managers, or of investing insufficient attention and resources in maintenance ('the problem must be yours').

As I have indicated above, the distinctive differences between German and Canadian practices were most frequently comprehended and described by those interviewed as arising from cultural dissimilarities. Indeed, the two sets of characteristic practices, expectations, attitudes and norms documented above might well be viewed as distinct industrial and business cultures. However, this diagnosis begs the obvious question, namely: how are such differences produced? More to the point, if one accepts that 'culture' (industrial or otherwise) is not some natural, prior, unchanging, and inherited whole, then how might it be socially produced and reproduced? One way of answering this is to set these cultural characteristics within their broader social and political context, by examining their relationship to concrete institutional and regulatory features. Given that many aspects of this context *also* differ markedly between Germany and the Anglo-American economies (*see* Christopherson, 1993; Keck, 1993; Goodhart, 1994; Herrigel, 1994; Wever, 1995; Whitley, 1999; O'Sullivan, 2000) it should come as no surprise that these larger, background differences might play a role.

In fact, I would argue that the differences described above can be linked quite directly to the nature of social institutions which regulate capital markets and business finance, labor markets, labor relations, and the corporate governance practices of user firms. Beginning with the most obvious point, differences in time horizon between the producers' German and North American customers (manifest here in several ways, from different payback preferences to outlooks on maintenance and training) are likely to be strongly influenced by the well-known differences in financing systems. Canada and the United States, based on the classic Anglo-American system of public capital markets for equity investments, have created business environments in which there is a strong division between financial and industrial capital. Shareholders usually exert significant power, creating strong pressures to produce short-term returns on investment. In contrast, German businesses raise the bulk of their equity capital through private investments. In a system in which financial institutions and industrial firms are closely linked, and in which (as a result of the labor relations institutions described below) a broader array of stakeholders (including workers and unions) are routinely represented on boards of directors, investment objectives are longer-term.

The pursuit of short-run returns is tempered by sources of capital which are patient or 'quiet', and by a stronger voice in favor of social returns, resulting from a system of corporate governance which provides for the direct representation of workers on the managing boards of many larger German firms. As a consequence, German industrial firms have considerably more latitude to wait longer periods of time for investments to bear fruit, explaining their apparent lack of interest in payback issues, relative to their North American counterparts.

On labor and training issues, here too there are sharp distinctions between the German and North American institutions and systems of regulation. One of the most distinctive features of the German economy is its system of labor relations based on the principle of 'co-determination'. Under this system, workers – both directly through firm-based works councils and indirectly through national unions – have a significant and institutionalized role in many aspects of the firm's decision-making, including training, technology acquisition and implementation, and day-to-day operations. Furthermore, and as a result of labor's institutionalized power, there are serious curbs on employers' ability to fire or lay off workers. Instead, the system works to encourage a stable employment relation characterized by lengthy employment tenure and the active use of internal labor market practices to manage firms' personnel needs. Furthermore, with a much greater degree of centralization of wage determination, and strong concordance between wages in union and non-union workplaces, interfirm competition based on wages is held in check.

All of this stands in sharp contrast to the Anglo-American norm, where employment relations are far less stable over the long term, where employers make far more extensive use of external labor market practices (hiring and firing), leading to the high turnover rates discussed earlier. Furthermore, apart from some key sectors such as automotive assembly, unionization rates are low and (at least in the United States) declining. As a result, the degree of interfirm variation in wages and working conditions is significantly greater than in Germany, and employers are encouraged to view labor cost as one of the chief dimensions of interfirm competition.

These fundamental differences in the institutional and regulatory framework surrounding employment play a large role in producing many of the practices and attitudes documented earlier and described so frequently as being cultural in origin. Most obviously, the instability in North American users' labor forces, noted earlier, can be seen as flowing directly from this institutional setting. So, too, can all of the consequences for training practices described above. In a system in which worker turnover is high, it makes perfect sense for Canadian employers to expect machine vendors or the state to take on the chief training role. Furthermore, when wage competition remains an all-too-tempting option, the skills of the workforce quickly become secondary. High turnover rates and politically weak labor also help explain the North American users' noted preference for 'technology' as something explicit, codified and embodied in inanimate machines rather than being socially produced through the joint action of skilled workers and machinery, in a stable relationship over time. They also provide an explanation for the North American users' much greater interest in exhaustive written manuals, which constitute an important part of the collected knowledge (in externalized rather than internalized form) about machine operation. Where workers are, effectively, interchangeable, this also is entirely understandable.

Likewise, the stark differences in maintenance practices can now be understood as arising, at least in part, from the structure of industrial investment finance and the institutions shaping capital markets. When investment capital is acquired on terms that are so strongly skewed in favor of quarterly returns, it should come as no surprise that Canadian (or American) users treat their capital equipment in a manner consistent with the prevailing truncated time horizons. When their decision-making horizon stops at two to three years, and their expectation is that a machine will be in active service only this long, it is understandable that managers will undervalue regular expenditures for the purpose of longer-term machine and system maintenance.

By the same token, it should not be surprising that North American workers do not develop the same sense of 'ownership' of their machinery as was seen to be the case in Germany, and do not engage in the same kind of lavish maintenance behavior that the German producers so admired in the practices of their domestic customers. Furthermore, when you have a system in which machine operators are much more likely to participate in the decision to purchase the machinery in the first place (including the process of deciding on technical specifications), this is a powerful force in the development of the sense of 'ownership' of a machine that was referred to earlier. This interaction becomes a vital conduit through which tacit knowledge about production process design flows directly between worker-operators and machinery producers. This crucially important participation by workers is itself the product of the German systems of industrial relations and corporate governance which actively accommodate worker input in technology implementation decisions, both directly and through their membership in works councils.

The differing perceptions of machinery complexity and ease of use are also understandable in these terms. It is clear that the entire constellation of regulatory features and rules in German industry foster a set of incentives and imperatives for a very different type of competitive outlook by its users of advanced technologies. When wages are removed from competition, when unions are strong, and more centralized bargaining systems ensure a high prevailing wage rate, and when layoffs are discouraged by labor market institutions, firms naturally turn to other means of competition – especially the technological capabilities or qualitative aspects of their products (Streeck, 1985). Thus, the allegedly cultural foundations of the German machinery builders' penchant for designing technically superior or 'overengineered' production systems should instead be understood as a rational response to such a competitive regime. So, too, does it become clear why their German customers not only demand such process technologies, but also invest heavily in the kinds of practices required to make these systems function at the peak of their potential capabilities. And in a setting where labor market institutions promote stability and minimize turnover, there are equally powerful incentives for businesses large and small to support the much-admired, highly developed national system of technical education, training and apprenticeship. Further, when competition is based less on prices than on technical and qualitative aspects of products, worker involvement becomes a key factor in the firm's success, including its success in implementing advanced process technologies (*see* Gordon, 1989 for an earlier analysis reaching similar conclusions).

Small wonder then, that German machinery has proven to be so difficult to implement with the same degree of effectiveness in North American plants. In the absence of a supportive social and institutional matrix – indeed, in the midst of what is clearly an antithetical regulatory regime – it should come as no surprise that Ontario users encountered such difficulties and frustrations (*see* Gertler, 1993 for a review of similar problems in other Anglo-American economies such as the United Kingdom).

Implications for regional innovation policy

This case study holds a number of important implications for our understanding of the technology transfer process (both *intra-* and *inter-*national), and also for regional policies aimed at promoting innovation, social learning and the dissemination of more advanced production methods.

First, it is clear from the experiences described above that the advanced machinery producers of Germany share much more than physical proximity and cultural (socio-linguistic) affinity with their customers in Germany. Underpinning their close, interaction-intensive relations with these domestic user firms (a relationship which supports learning based on tacit knowledge flow) is a shared macro-regulatory framework. This framework is reflected especially in capital markets, labor markets, corporate governance and industrial relations, and it helps shape a shared set of expectations, practices, attitudes and norms which facilitate and enable social learning processes. Taken together, they resemble a set of common rules of behavior which we might view as constituting a distinct 'industrial culture'; nevertheless, they are also clearly influenced by a set of overarching economic institutions. Moreover, the fact that the German firms interviewed in the study came from several different regions of Germany and yet reported very similar sources of problems and conflict with their North American customers, underscores the point that *national* institutions play a key role in creating the enabling conditions for regionally bound social learning.

Second, and following directly from the first point, while 'close', extended interaction between user and producer is indeed a key ingredient of successful technology transfer, it ought to be regarded as a necessary but not sufficient condition for a satisfactory outcome from this relationship. In this sense, consumers of German-built advanced machinery in Canada and the United States face a double disadvantage: they are physically distant from the geographical source of their machinery (which makes the required face-to-face interaction considerably more expensive), and they are also institutionally distant from the producers of their process technologies.

Third, the more general lesson arising from this insight is that whenever user firms invest in 'leading edge', 'best practice' or 'state-of-the-art' technologies which originate from sources that are institutionally distant, they can expect serious implementation difficulties to arise. This finding has particular significance for the recent generation of public programs aimed at 'modernizing' manufacturing by encouraging firms to adopt new advanced process technologies.

There is little doubt that such forms of intervention perform a useful role in promoting the dissemination of state-of-the-art technologies (although, as Gittell *et al.*, 1996 point out, the past few years have seen a lively debate unfolding in the United States over the degree to which such benefits outweigh the costs associated with program delivery). As many of these technologies represent a fairly distinct departure from past practices, and are not easily incorporated into existing operations, the 'assisted' model of technology transfer (either through centers or extension agents) is likely to promote a faster rate of uptake and, hopefully, more effective use, than would unfold through the unassisted workings of purely market-based diffusion processes. Indeed, the fact that even technology-rich manufacturing nations such as Japan and Germany have developed their own extensive networks of technology transfer centers (Shapira, 1996; Cooke and Morgan, 1990) suggests that this is a vital and useful form of intervention. In short, the relatively unsophisticated user firm is likely to be better off when such assistance is available than when it is not.

However, the case study documented in this chapter holds important implications for the well-meaning industrial extension agent or the staff of a regional technology transfer center. It is simply not enough to make their client firms aware of the existence and source of the 'best available' or 'best practice' technologies, or even to help their clients come to an agreement with the provider of such technologies. Indeed, by doing so, they may in fact be doing their clients a disservice if they have inadvertently encouraged them to invest in process technologies designed for application in workplaces shaped by a very different social and economic context (*see* Gertler, 2001).

A fourth policy implication arising from the analysis in this chapter is that there may well be significant benefits for both parties when users and producers of advanced process technologies are 'close' to one another (in both the physical and the regulatory/institutional sense). Putting this another way, it seems clear that the technology transfer process is indeed strongly subject to spatial limits, and that the positive externalities arising from interfirm interaction are also geographically bound. Hence, there may be significant advantages in the form of social learning opportunities arising from the promotion of a home-grown, domestically based machinery-producing sector – a lesson not lost on the US producers of semiconductors, who benefited tremendously from the Sematech initiative to boost the position of the American producers of semiconductor manufacturing equipment (*see* Stowsky, 1987; Tyson, 1992; Porter, 2000). As the governments of nations such as Japan, South Korea and Taiwan have discovered, there are indeed good reasons (beyond those related to international security and strategic positioning) to promote the development of home-grown process technologies in sectors related to one's population of user firms (Gertler *et al.*, 1995).

Thus far, all of the implications discussed relate to the transfer of 'technology' in the narrowest sense of the term. Yet, recalling our earlier conclusion that technology also embraces the social organization of work, including the nature of external relations between transacting firms, it is important to reflect on what implications the case study might hold for similarly well-meaning attempts to institute new types of relations between firms in a given region.

Collaboration culture: the production of social capital and constraints on social learning

Our earlier discussion has endeavored to show how a set of practices implemented *inside* individual workplaces do not arise from some inherited, static business culture, but are instead strongly influenced by a set of concrete institutions. It should be pointed out, however, that the same kind of analysis may be fruitfully applied to relations *between* firms and the ease with which they can engage in social learning processes. As was noted in the introductory sections of this paper, much has been written about the region-specific cultures that are said to foster (in some regions) trust-building, openness, cooperation, information-sharing, and interfirm learning, leading to more successful innovation and production.

However, without wishing to deny the very special character of the social learning dynamics in the innovative industrial districts of Italy, Germany and elsewhere, it is important to appreciate the role played by supraregional (largely national) institutions in creating a supportive regulatory context to facilitate the proliferation of extra-market (that is, untraded) interaction and collaboration. In this sense, the national regulatory framework can be seen as providing the necessary but not sufficient conditions for local interfirm learning and related practices to thrive – the bedrock upon which social capital is constructed, if you will. By the same token, the absence of a conducive regulatory framework might well make it difficult for such social learning relations to develop and survive. Several examples serve to illustrate these points more concretely.

First, returning to the case of Germany, the forms of interfirm cooperation which abound in Baden-Württemberg have been enabled by a sympathetic set of national regulatory features – particularly in the realm of labor markets and industrial relations. These features, already described in earlier sections of this paper, have the net effect of largely removing from the field of competition such potentially divisive issues as wage determination, working conditions, and worker empowerment, as these are set through a combination of national and regional industry-wide agreements which are frequently extended to cover even non-union employers. As a result, firms resort to competing largely on the basis of product and process technologies and quality.

The impact on interfirm relations is distinctive and unmistakable. On the one hand, it produces a fierce competitiveness and strong degree of secrecy which discourage direct collaboration with one's competitors. After all, if one's technology is the primary source of competitive advantage, then this will be jealously guarded and firms will go to extraordinary lengths to keep proprietary technical information from their competitors (Cooke and Morgan, 1990; Cooke, Morgan and Price, 1993; Gertler, 1996). At the same time, in pursuit of technical excellence, firms are encouraged to cooperate vertically with specialized suppliers of key inputs and, because of minimal interfirm variation in wages, such outsourcing practices do not seduce firms to engage in wage-lowering competition, nor do they undermine the power of unions.

Furthermore, the constellation of nationally legislated labor market institutions do facilitate considerable *indirect* cooperation between firms in the same industry, through industry-based employers associations and chambers of commerce in which membership is mandatory. These institutions play a major role in designing and delivering training programs, and in spreading common organizational innovations throughout entire industries. As Wever (1995, p. 11) notes, the larger national structures which give rise to such institutions in Germany 'create incentives that support the sharing of information and knowledge and thus encourage the diffusion of organizational innovation'. Furthermore, she sees similar structural conditions operating within the Japanese economy (p. 11): 'The shared logic underlying the organization of the German and Japanese political economies is not cultural. It is the logic of institutional and organizational linkages.'

On the Japanese economy, widespread interfirm cooperation, including 'relational contracting' between buyers and suppliers, has received considerable attention in the business press. In assessing the literature on this phenomenon (and especially the role of supplier associations in promoting interfirm learning and diffusion of 'best practice'), Morgan (1996) has concluded that 'culturalist' interpretations of this set of practices are inadequate. Instead, 'trust and reciprocity (that is, social capital), far from being pre-existing cultural assets, hardly existed prior to the war economy' (p. 4). Drawing on the work of Nishiguchi (1994), he continues:

> In fact such social capital was actively constructed through a combination of corporate necessity (like the fact that large firms had to delegate tasks to suppliers to meet surging post-war demand) and legislation which promoted small firm associations on the one hand and prevented unfair subcontracting practices on the other.

As a result, interfirm relations in Japan became far less exploitive. With the benefit of new rules governing interfirm subcontracting, both supplier and buyer came to benefit from a social process of joint problem-solving.

It seems plausible, based on the two cases described above, that the macroregulatory framework is just as capable of shaping interfirm behavior as it is of shaping intrafirm practices. Hence, some obvious implications flow from this, which regional innovation-promoting agencies should heed.

First, it seems clear that if collaborative relations between firms are not underpinned by a 'cooperation-friendly' macroregulatory framework, then the efforts expended by regional agencies to promote innovation, technology transfer and network-building to broker cooperation, information-sharing, and the development of trust between individual firms will be largely wasted – at least, so long as the goal of policy is to build 'European-style' social capital and cooperative behavior (Cohen and Fields, 1999). It is clearly inadequate to try to convince firms to cooperate with one another simply by exhorting them to do so. Similarly, leading field trips to Italy, Germany, Denmark, and other stops on the now well-worn world tour, in order to show firm managers what joyous benefits await them once they adopt a new cooperative stance, will lead to little of lasting benefit, as such initiatives tend to dwell solely on the actions of firms, while failing to explore the larger institutional matrix that enables and facilitates such actions.

Even more sophisticated attempts to build a 'collaboration culture' and offer firms opportunities for learning-through-interacting by the promotion of research networks and consortia, sectoral centers for training and/or research, and other tangible elements of a regional innovation system may bring only limited results, if these, too, tend to ignore the role of the broader, systemic institutional framework in creating (or undermining) incentives to cooperate. As Wolfe and Gertler (1998; Gertler, Wolfe and Garkut, 1998; *see also* Wolfe, this volume) have shown, such policy experiments in Ontario during the first half of the 1990s were ultimately frustrated by systemic impediments to cooperation. Hence, so long as the workings of labor markets are regulated in a way that discourages investment in training by firms, the resultant poaching of skilled labor and managerial talent will do little to kindle the spirit of cooperation between firms competing for the same skilled labor. So long as capital market structures and mechanisms for industrial finance impose a strongly short-term outlook on firms, they will be understandably reluctant to invest in the creation and maintenance of long-term interfirm relations.

Nevertheless, notwithstanding the evidence provided in this chapter, I do not wish to argue that regional institutions are unimportant in the creation of conditions conducive to innovation, adoption and diffusion of progressive industrial practices and even social learning. They are crucial, for example, in helping us understand the very distinctive trajectories and 'cultures' of different regions within the same nation-state (for example, Germany's Ruhrgebiet compared to Baden-Württemberg, or Silicon Valley versus Route 128 in the United States). Thus, regional institutions *do* indeed matter. But to emphasize the role of local or regional institutions to the exclusion of all others – or to fail to consider how these two scales of regulation interact to produce outcomes in particular places – is as misleading as denying any role for regional institutions at all.

Hence, it remains important for scholars and policy-makers to appreciate the importance of national institutions in creating the enabling, accommodative space within which particular regional growth phenomena may arise. In this sense, then, we can understand the spatial construction of social learning dynamics, including those which socially 'produce' manufacturing technologies, as occurring through the interaction of national and subnational regulatory forces.

Note

1   The author gratefully acknowledges the Social Sciences and Humanities Research Council of Canada for its support of the research on which this chapter is based. He would also like to acknowledge David Wolfe and other colleagues in Canada's Innovation Systems Research Network for many helpful conversations and insights.

References

Best, M. (1990) *The New Competition: Institutions of Industrial Restructuring*, Cambridge, MA: Harvard University Press.

Bosworth, B. and S. Rosenfeld (1993) *Significant Others: Exploring the Potential of Manufacturing Networks*, Chapel Hill, NC: Regional Technology Strategies, Inc.

Braczyk, H-J., P. Cooke, and M. Heidenreich (1998) *Regional Innovation Systems*, London: UCL Press.

Britton, J.N.H. and J.M. Gilmour (1978) *The Weakest Link: a Technological Perspective on Canadian Industrial Underdevelopment*, Ottawa: Science Council of Canada, Background Study No. 43.

Christopherson, S. (1993) 'Market Rules and Territorial Outcomes: the Case of the United States', *International Journal of Urban and Regional Research*, 17, 274–88.

Cohen, S. and G. Fields (1999) 'Social Capital and Capital Gains in Silicon Valley', *California Management Review* 41, 108–30.

Cooke, P. (1999) 'The Co-operative Advantage of Regions', in T.J. Barnes and M.S. Gertler (eds) *The New Industrial Geography: Regions, Regulation and Institutions*, London: Routledge, pp. 54–73.

Cooke, P. and K. Morgan (1990) *Industry, Training and Technology Transfer: the Baden-Württemberg System in Perspective*, Cardiff: Regional Industrial Research Paper No. 6, University of Wales.

Cooke, P. and K. Morgan (1998) *The Associational Economy*, Oxford: Oxford University Press.

Cooke, P., K. Morgan, and A. Price (1993) *The Future of the Mittelstand: Collaboration versus Competition*, Cardiff: Regional Industrial Research Paper No. 13, University of Wales.

Coopers and Lybrand (1994) *Good Practice in Managing Technology Transfer Networks: Ten Years of Experience in the SPRINT Program*, 2 vols, Brussels: European Commission, CEC-DGXIII/D/4.

Dosi, G. (1988) 'Sources, Procedures and Microeconomic Effects of Innovation', *Journal of Economic Literature*, 26, 1120–71.

Florida, R. (1995) 'Toward the Learning Region', *Futures* 27, 527–36.

Gertler, M.S. (1993) 'Implementing Advanced Manufacturing Technologies in Mature Industrial Regions: towards a social model of technology production', *Regional Studies*, 27, 259–78.

—— (1995) ' "Being There": Proximity, Organization, and Culture in the Development and Adoption of Advanced Manufacturing Technologies', *Economic Geography*, 71, 1–26.

—— (1996) 'Worlds Apart: the Changing Market Geography of the German Machinery Industry', *Small Business Economics*, 8, 87–106.

—— (2001) 'Best Practice? Geography, Learning, and the Institutional Limits to Strong Convergence', *Journal of Economic Geography*, 1, 5–26.

——, S. DiGiovanna and D. Tassie (1995) *The Structure and Strategic Importance of the Machinery, Tool, Die and Mould Sector in Ontario*. Report prepared for the Ontario Machinery, Tool, Die and Mould Sectoral Partnership Project, Ministry of Economic Development and Trade, Toronto.

——, D.A. Wolfe and D. Garkut (1998) 'The Dynamics of Regional Innovation in Ontario', in J. de la Mothe and G. Paquet (eds.) *Local and Regional Systems of Innovation*, Amsterdam: Kluwer Academic Publishers, pp. 211–38.

Gittell, R., A. Kaufman and M. Merenda (1996) 'Rationalizing State Economic Development', In U. Staber, N. Schaefer and B. Sharma (eds) *Business Networks: Prospects for Regional Development*, Berlin: Walter de Gruyter, pp. 65–81.

Goodhart, D. (1994) *The Reshaping of the German Social Market*, London: Institute for Public Policy Research.

Gordon, R. (1989) 'Beyond Entrepreneurialism and Hierarchy: the Changing Social and Spatial Organization of Innovation'. Paper presented at the Third International Workshop on Innovation, Technological Change and Spatial Impacts, Selwyn College, Cambridge, UK, 3–5 September.

Herrigel, G. (1994) 'Industry as a Form of Order: a Comparison of the Historical Development of the Machine Tool Industries in the United States and Germany', in J.R. Hollingsworth, P.C. Schmitter and W. Streeck (eds) *Governing Capitalist Economies: Performance and Control of Economic Sectors*, New York: Oxford University Press, pp. 97–128.

Keck, O. (1993) 'The National System for Technical Innovation in Germany', in R. R. Nelson (ed.) *National Innovation Systems: a Comparative Analysis*, New York: Oxford University Press, pp. 115–57.

Leadbeater, C. (1999) *Living on Thin Air*, London: Viking.

Lundvall, B-Å. (1988) 'Innovation as an Interactive Process: from User–Producer Interaction to the National System of Innovation', in G. Dosi, C. Freeman, G. Silverberg and L. Soete (eds) *Technical Change and Economic Theory*, London: Frances Pinter, pp. 349–69.

Lundvall, B-Å. and B. Johnson (1994) The Learning Economy, *Journal of Industry Studies*, 1, 23–42.

Morgan, K. (1996) 'Learning-by-interacting: Inter-firm Networks and Enterprise Support', in *Local Systems of Small Firms and Job Creation*, Paris: OECD.

—— (1997) 'The Learning Region: Institutions, Innovation and Regional Renewal', *Regional Studies* 31, 491–503.

Nelson, R.R. (ed.) (1993) *National Innovation Systems: a Comparative Analysis*, New York: Oxford University Press.

Nishiguchi, T. (1994) *Strategic Industrial Sourcing: the Japanese Advantage*, Oxford: Oxford University Press.

Nonaka, I. and H. Takeuchi (1995) *The Knowledge-Creating Company: How Japanese Companies Create the Dynamics of Innovation*, Oxford: Oxford University Press.

O'Sullivan, M. (2000) *Contests for Corporate Control: Corporate Governance in the United States and Germany*, Oxford: Oxford University Press.

Polanyi, K. (1944) *The Great Transformation*, Boston: Beacon Press.

Porter, M.E. (1991) *Canada at the Crossroads*, Ottawa: Business Council on National Issues and Minister of Supply and Services.

—— (2000) 'Locations, Clusters, and Company Strategy', in G.L. Clark, M.P. Feldman and M. S. Gertler (eds) *The Oxford Handbook of Economic Geography*, Oxford: Oxford University Press, pp. 253–74.

Putnam, R. (1993) *Making Democracy Work*, Princeton, NJ: Princeton University Press.

Sabel, C. (1989) 'Flexible Specialization and the Resurgence of Regional Economies', in P. Hirst and J. Zeitlin (eds) *Reversing Industrial Decline?* Oxford: Berg, pp. 17–70.

Sabel, C. (1992) 'Studied Trust: Building New Forms of Co-operation in a Volatile Economy', in F. Pyke and W. Sengenberger (eds) *Industrial Districts and Local Economic Regeneration*, Geneva: International Institute of Labour Studies, pp. 215–50.

Saxenian, A. (1994) *Regional Advantage: Culture and Competition in Silicon Valley and Route 128*, Cambridge, MA: Harvard University Press.

Sayer, A. and R. Walker (1992) *The New Social Economy: Reworking the Division of Labor*, Oxford: Basil Blackwell.

Scott, A.J. (1996) 'Regional Motors of the Global Economy', *Futures*, 28, 391–411.

—— (2000) 'Economic Geography: the Great Half-century', in G.L. Clark, M.P. Feldman and M.S. Gertler (eds) *The Oxford Handbook of Economic Geography*, Oxford: Oxford University Press, pp. 18–44.

Shapira, P. (1990) *Modernizing Manufacturing: New Policies to Build Industrial Extension Services*, Washington, DC: Economic Policy Institute.

—— (1996) 'Modernizing Small Manufacturers in the United States and Japan: Public Technological Infrastructures and Strategies', in M. Teubal, D. Foray, M. Justman and E. Zuscovitch (eds) *Technological Infrastructure Policy: an International Perspective*, Dordrecht: Kluwer Academic Publishers, pp. 285–334.

Shapira, P., J. Youtie and J.D. Roessner (1996) 'Current Practices in the Evaluation of US Industrial Modernization Programs', *Research Policy*, 25, 185–214.

Staber, U., N. Schaefer and B. Sharma (eds) (1996) *Business Networks: Prospects for Regional Development*, Berlin: Walter de Gruyter.

Storper, M. (1992) 'The Limits to Globalization: Technology Districts and International Trade', *Economic Geography*, 68, 60–93.

—— (1995) 'The Resurgence of Regional Economies, Ten Years Later', *European Urban and Regional Studies* 2, 191–221.

Stowsky, J. (1987) 'The Weakest Link: Semiconductor Production Equipment, Linkages, and the Limits to International Trade'. Working Paper 27, Berkeley Roundtable on the International Economy, University of California at Berkeley.

Streeck, W. (1985) 'Industrial Relations and Technical Change in the British, Italian, and German Automobile Industry: three case studies'. Discussion Paper IIM-LMP 83–5, Berlin: Wissenschaftzentrum.

Tyson, L. (1992) *Who's Bashing Whom? Trade Conflict in High-Technology Industries*, Washington, DC: Institute for International Economics.

Wever, K.S. (1995) *Negotiating Competitiveness: Employment Relations and Organizational Innovation in Germany and the United States*, Boston: Harvard Business School Press.

Whitley, R. (1999) *Divergent Capitalisms: the Social Structuring and Change of Business Systems*, Oxford: Oxford University Press.

Wolfe, D.A. and M.S. Gertler (1998) 'Ontario's Regional System of Innovation', in H-J. Braczyk, P. Cooke and M. Heidenreich (eds) *Regional Innovation Systems*, London: UCL Press, pp. 99–135.

7
Institutions of the Learning Economy[1]
Michael Storper

Competitiveness and the learning economy

Theories of competitiveness abound today, as do descriptive monikers for the new economy: post-industrialism, the informational economy, the knowledge-based economy, flexible specialization, post-Fordism. Though each of these labels helps in understanding some dimensions of contemporary economic activity, the logic of the most advanced forms of economic competition – those capable of generating high-wage employment – can best be described as that of learning. Those firms, sectors, regions and nations that can learn faster or better (higher quality or cheaper for a given quality) become competitive because their knowledge is scarce and therefore cannot be immediately imitated by new entrants or transferred, via codified and formal channels, to competitor firms, regions, or nations. The price-cost margin of such activities can rise, while market shares increase; the resulting rents can alleviate downward wage pressure. In this respect, such activities are promising for high wage areas. But the key defining condition of this happy picture is that these activities are only temporarily immune to relocation or to substitution by competitors. Economies must therefore be equipped to keep outrunning the powerful forces of standardization and imitation in the world economy. Once they are imitated or their outputs standardized, then there are downward wage and employment pressures. They must become moving targets by continuing to learn. Learning may enhance product differentiation at any given moment, or it may take the principal form of constantly adapting the configuration of products and processes so as to anticipate the competition.

The appellation 'learning economy' has considerable and important differences with other concepts applied to the 'new economy' of the post-1970 period. Its central emphasis is on *time* in sustaining a desirable form of imperfect competition, characterized by ongoing product-based learning. It generates temporary non-substitutability (scarcity) of key inputs, especially labor and human relations. It should be stressed that the term learning as used henceforth refers specifically to product-based technological learning (PBTL). This definition stresses the usefulness of technological change in adapting the product, which is the principal vector of competition. Learning may enhance product differentiation at any given moment, or it may take the principal form of constantly adapting the configuration of products and processes so as to *anticipate* the competition. Such product adaptation may involve many upstream forms of technological change, but in and of themselves, unsystematic forms of technological change will not be adequate to generate competitiveness. PBTL is quite different – analytically speaking – from technological imitation in production processes (such modernization being the main subject of management literatures, even those that now use fashionable terms such as 'learning', 'knowledge', or 'innovation').

The notion of a PBTL economy differs from a number of other current concepts as well. 'Informational' economy analyses suggest the existence of a new form of competition based on the collection and transmission of information, but they suffer from imprecision. Standardized and codified information and the high-tech processes and technologies used to transmit and analyze it are not the essence of contemporary economic competition: true knowledge of intellectual skill is at the heart of this process. The notion of a 'knowledge-based' economy comes closer to the target, but it also has a certain analytical imprecision, in that it fails to distinguish between knowledge as a stock – a factor of production – and as a flow – a subject of competition, which is the center of the creative and allocative dynamics of capitalism. Learning is the center of this knowledge-based competitive process and it includes not only intellectual (technical and scientific) knowledge, but also manual and implicit knowledge, and tools and organizations that accompany both of them.

'Learning' is also a more effective theoretical notion than the label 'post-Fordist', in that it affirms what is central to contemporary economic competition: not simply that it comes after the complete stability allegedly associated with mass production. It is, rather, that the nature of economic fluctuations has changed, from predictable risks to true uncertainty, and the latter is not amenable to the forms of production smoothing which were described by Galbraith and Chandler in their analyses of postwar mass production. It is thus more general than any single 'best practice' image of the contemporary economy; the particular practices that facilitate learning for competitiveness vary widely according to sector, country, and institutional context.

Finally, the notion of a learning economy rejects the central arguments of post-industrialism: learning concerns manufacturing, which continues to matter, as well as services. Learning can concern low-technology industries, which can generate high-wage jobs, as well as high-technology sectors. Moreover, all are organized in fundamentally 'industrial' ways (wage labor, capital, produced inputs, search for efficiency through divisions of labor, and so on). The industrial way of doing things, however, has entered a new phase where its content is being reshaped by the advent of learning-based competition.

PBTL activities, while central to the direct objective of generating high-wage, high-skill 'knowledge-intensive' employment, are never going to account for the majority of output or employment in any given economy. Other kinds of activities (locally serving activities, scale-based production, and so on) will continue to embody high proportions of employment. But PBTL has propulsive effects on economies in a number of ways: technological spillover effects can widen and lengthen the wealth-producing properties of learning (both upstream and downstream, and horizontally into technologically complementary activities), while the quasi-rents earned from imperfect competition can be channelled through the producing economy in the form of wages, investment and cumulative advantage.

To say that the learning economy is necessary to high-wage employment generation is not to claim that it represents a complete economic strategy. All the traditional tasks also remain necessary: balancing production and consumption; finding the right mix between export-oriented and locally serving activity; ongoing productivity improvements; and balanced reallocation of labor. But these traditional tasks of long-term economic management are by themselves no longer sufficient to generate adequate quantities of high-quality employment. That is the role of the learning economy.

The organization of the learning economy: conventions

Coordination as the key problem of economic life and convention as a key form of coordination

Coordination among persons is the central problem of economic life. All productive activity depends on reciprocal, coherently matched actions by others which, if not forthcoming, will render our own actions inefficient or unproductive. Yet virtually all such situations of action are beset by uncertainty – each of us faces uncertainty in deciding what we should do with respect to a given set of circumstances. Part of this uncertainty is 'secondary', that is, it comes from the fact that others upon whom we depend also face uncertainty on their side, so they do not know, with assurance, what they will do; part of it comes from imperfect knowledge or communication of their intentions. All this is another way of saying that productive activity is, of necessity, a form of collective action founded on the paradox of individual actions. The question is how actors manage to get themselves into successfully coordinated forms of collective action.

The perspective which has been developed over recent years known as the 'social science of conventions' holds that coordination is achieved when actors develop appropriate conventions for doing so. The latter may be defined to include taken-for-granted mutually coherent expectations and practices, which are sometimes manifested as formal institutions and rules, often not. There are many different conventions in productive activity, for two principal reasons: because uncertainty, though pervasive, takes different forms according to the many and varied products of the economy, in light of the production technologies, markets, and resources associated with different kinds of products; and because historically constituted and geographically differentiated groups of actors bring to bear different rationalities to the uncertain situations of action they encounter.

Learning-based production systems are particular cases of conventional economies. They tend to involve particularly intense conventional underpinnings owing to the existence of rather 'non-cosmopolitan' forms of knowledge. These are the cognitive foundations of learning and their effects on economic transactions. In addition, high rates of learning occur only when there are high levels of overall coherence of the many and varied conventions involved in a production system. Each of these will now be discussed in turn.

Transactions and cognition, relations and conventions

Transactions between firms and the external environment – labor markets and information-rich institutions, such as universities, trade associations and other institutions – are particularly important to learning. It is to the substantive content of transactions and their 'governance' or 'regulation' that we must look in order to penetrate deeper into the learning process itself. Yet transactions inherently involve uncertainty. Market systems are said to compensate uncertainty via the law of large numbers (with highly substitutable products or factors),[2] but this only really works when competition is perfect. In the case of learning-oriented production systems, the relations which compose the system cannot be of such large numbers and so substitutable that they function like true markets. Unlike transactions of standardized and substitutable goods, factor inputs, and information, transactions associated with the kind of learning analyzed here involve the development and – perhaps even more important – the mutually consistent interpretation of information which is not fully codified, hence not fully capable of being transmitted, understood, and utilized independently of the actual agents who are developing and using it. Stated another way, when the knowledge or skills involved have cognitive dimensions which are non-cosmopolitan (that is, non-universalist, particularistic), the transaction becomes a concrete, real *relation* and has dimensions which cannot be abstracted and thus divorced from its existence as a relation.

Moreover, beyond such tension between non-cosmopolitan technical models and meta-modeling, every kind of production system has to cope with some form of fluctuations in markets, product design, available technology and prices, which make difficult the full cognitive routinization of relations between firms, their environments and employees. Many such fluctuations, if they are to be dealt with in such a way that efficiency losses or conflicts are to be avoided, involve less than bureaucratic procedures and adjustment mechanisms, which are highly embedded in informal or partial-ly formal rules and practices.

There are two levels of this relational quality of transactions. In the first, personal contacts, knowledge of the other, and reputation are the basis of the relation (Asanuma, 1989). In many other cases, however, transactions are not so completely idiosyncratic; they do have dimensions which can be repro-duced or imitated by other agents. Most conventions are a kind of half-way house between fully personalized and idiosyncratic relations and fully deper-sonalized, easy to imitate relations (although even the latter do have conven-tional foundations, not natural or behaviorally universal foundations).

Conventional or relational transactions (henceforth C-R) affect many dimensions of production systems, but the nature and functions of such conventions differ from industry to industry, according to the nature of the product, the economic fluctuations associated with its markets and produc-tion processes, and the type of learning which is possible.[3] C-R transactions may be found in at least five principal domains: (a) interfirm 'hard' transac-tions, as in buyer–seller relations that involve market imperfections; (b) interfirm 'soft' transactions, as in the diffusion of non-traded informa-tion about the environment or about learning, for example through circula-tion of personnel through the same external labor market or through contact between producers; (c) in hard and soft intrafirm relations, as the bases for the functioning of large firms which are 'internally externalized' in the way we noted above; (d) in factor markets, especially labor markets, which involve skills that are not entirely substitutable on an interindustry or inter-regional basis, i.e., where there are industry- or region-specific dimen-sions to workers' skills; and (e) in economy-formal institution relationships, where universities, governments, industry associations and firms are only able to communicate and coordinate their interactions by using channels with a strong relational-conventional content.

Note that in this analysis, the learning economy and its conventional-relational foundations is not based on a stark contrast between internal ownership and externalization of production systems, or on hierarchies ver-sus markets or external-embedded networks, but rather on the notion that all advanced, learning-based forms of economic activity involve complex transactional structures which in turn have a high conventional-relational content. Indeed, in many respects,[4] the big firm *is* a set of relations and it is such relations, as much as the physical or financial capital the firm com-mands, which allow it to compete and survive.

Building relations: precedent, talk and confidence

Relations and conventions are recursive outcomes of precedents which act as guides on action, and are reinterpreted and re-evaluated for their efficacy, and reproduced as conventions when they work to coordinate action under conditions of uncertainty. The problem is that if such precedents do not exist or are not adequate to the kind of learning system which is to be created, deliberate institutions to create them suffer from the circularity identified above. And a learning system is a complex organizational structure with many different actors and transactions between them, hence many different conventions and types of relations, built on precedents which are effectively indivisible, if the learning system is to work. It is probably no accident that considerable recent research reveals the cardinal importance of 'soft' factors such as 'civic culture' (Putnam, 1992; Doeringer and Terkla, 1990) in the performance of democratic institutions, but that few venture any policy-oriented recommendations on how the lack of such a culture could be redressed. Very unorthodox policy strategies are needed in order to break out of these labyrinthine prisons. Two of these may be labeled, respectively, 'talk' and 'confidence'.

The circular relation between public institutions and the institutionalized learning economy requires that the parties to public institutions somehow be convinced of the utility of having a public institution help in supporting the conventions and relations which make up the institutionalized learning economy. That is, they must share a convention of the utility of the public institution in some specific domain, before it can even get started. *Talk* between the parties may be one approach. Much has been said about the difference between institutions that function via a combination of loyalty and voice, versus those that rely on exit for adjustment and structure (Hirschman, 1970). Talk is upstream of voice, in that there is no institution yet existing in which the channels for voice among loyal parties are already established.

Talk refers to communicative interaction, designed not simply to transmit information and relay preferences, but to achieve mutual understanding.[5] In the case of prospective learning, information from other experiences where learning has worked (on evolutions in product markets, on suggested potentials for the parties at hand, given their current resources and skills) can be valuable as a stimulus, even though it cannot be represented as 'experimental', that is, as automatically useful or valid in other circumstances. Such information can, however, be used as the valid pretext for talk.

It can immediately be objected that if there is no tradition of communication or, worse, if there is distrust or antipathy, what is the possible basis for talk? The objection is important: it is probably difficult to stimulate talk, precisely because talk is not free; it takes time and effort, and payoffs are not evident, especially if the history of relations is bad, or relations are satisfactory for those already in them (Hirschmann, 1970). On the other hand, talk is cheap; it is not that costly and the risks are relatively low. Public institutions thus certainly have a possibility of getting low-cost talk going. Talk alone, however, is unlikely to be sufficient if such fears exist; but rather than bribing the parties with incentives (or at least doing so in more than a temporary way), it would be better to offer them some sort of reinsurance (Sabel, 1993), a safety net (at least partial) for failure, forcing them to reveal the efficacy of talk and their propensity to have confidence. Finally, talk of the sort referred to here means intensive communicative interaction. Shallow contact will not do. Thus, such talk is more likely to get going if carried out in low-cost ways where *depth* is also possible (more on this shortly).

Precedents which underpin conventions or relations inherently involve confidence, without which single events would be just that, and would have no impact on future expectations. Insofar as conventions and relations involve expectations about how others will interact with you in situations which involve some uncertainty, such confidence involves a measure of vulnerability: it is necessary for interacting agents to place themselves in a position where, should the other not follow precedent, they will suffer a real loss (Lorenz, 1992). To have confidence in what others will do is, in this sense, to trust them, not in the metaphysical sense, but in the analytical sense of making oneself vulnerable, on the basis of confidence in the precedent. But how to establish such confidence so as to bring into being precedent, relation and convention, where it does not exist or worse, where there are histories of mistrust, broken promises, antagonisms?

Talk may involve the parties in getting the ball rolling on a learning project, but it does not establish confidence in the specific sense that generates precedent and convention. Bribery through special material incentives (subsidies, and so on) provided by a public institution to private actors is likely to work only as long as the incentives last; if all actors calculate that the other actors only do what they do because of special incentives, then a convention based on incentives is established and with it, the possibility of lock-in to subsidy. Therefore, if the intention of a policy is to establish learning conventions that are not dependent on permanent subsidies, other approaches will have to be tried, or early incentives will have to be slowly replaced with precedents in other, non-subsidized forms of making oneself vulnerable.

One method of creating confidence in a sea of non-confidence is, of course, bureaucracy (hierarchy).[6] It has been found, in economic policy-making, that certain projects are amenable to isolation from the overall economic culture, by internalizing them within hierarchical bureaucracies. The military is the model. Defense procurement in the USA, or major indivisible high-technology projects such as the French TGV, are carried out by quasi-military bureaucracies with strong financial incentives and command-and-control authority. But internalization is not a solution for much of the learning economy, precisely because of the open-endedness and high degree of risk of much learning, which nobody in society wants to pay to internalize, and where the technological character of the product does not permit near-monopoly (scale economies, extreme indivisibilities, as in the case of the TGV).[7] Some other method of building confidence must be used.

Small, repeated, experimental interactions may be useful for this purpose. Experiments, as a policy device, means actually setting the parties to work in limited relations which facilitate learning and attempting to build up in complexity. It does not mean trying to prove the utility of any general, abstract solution. Most importantly, such experiments must proceed 'as if' confidence existed. Small experiments build on the communicative understanding that comes from talk, asking the parties to interact by suspending their fears and doubts. The likelihood of getting the parties to act *as if* confidence existed, as the first step toward establishing real precedents, should logically rise with the degree of knowledge they have about each other. Depth is one dimension: the more I know about you in a specific domain; but breadth is another: the more I know about you in general, through collateral forms of information, the more I will be willing to enter into deep contact. These include risks of collective sanctions for violating the terms of a relation; reputation effects due to rich information flows; cultural proximity, behavioral norms which shape the anticipations of agents; and frequency of contacts in and outside of the particular business context (Haas-Lorenz, 1994). Attention has also been called to 'institutional thickness' – multiple, partially overlapping and partially redundant institutions – as a basis for breadth (Amin and Thrift, 1993).

Depth has a complicated geography, in that professional interactions, in some cases, have channels involving strong specific long-distance relations and weak local ones, above all in specialized or highly formalized (cosmopolitan) professions. Still, even in such circles, local relations often involve forms of depth not achieved in long-distance, infrequent contacts. Breadth has a more uniformly localist dimension: we are more likely to have information on someone's reputation, and to be able to validate it by interpreting it against a context with which we are intimately familiar, in a local context.[8] There is thus some relationship between localness and the mutual knowledge that should allow parties to act as if confidence existed, as a first step toward generating precedent. Talk and confidence – depth aided by confidence due to breadth – while not the province of the locality, are in some cases (certain products, certain worlds of production) more likely to succeed when they are geographically localized. This is not a hard-and-fast rule, of course, and much more theory and evidence, is needed before these relationships can be understood in a policy-relevant fashion.

The new heterodox policy framework

There are many intricate dimensions of talk and confidence-building as the vehicles for creating precedent, relation, and convention. Who should talk? What should they talk about? What techniques should be used to facilitate such talk? What small relations should be attempted first? What kind of encouragement should be offered to get the parties to suspend skepticism? The answers will vary not only according to the kind of world which talk is intended to get started, but also according to the starting point of the parties. Some very modest beginnings will be attempted in this section.

Starting points: strategic assessment

It has long been standard practice in industrial policies to carry out strategic assessments of local, regional or national possibilities (depending on the policy's target). The idea is to eliminate unreasonable goals by assessing the existing state of such factors as technological level, the labor market, infrastructure, market structure, and so on. Such analyses, in practice, vary greatly in quality, and unfortunately there is a high propensity for error, especially excessive optimism (since the assessments are usually paid for by agencies with a vested interest in being in the policy business). Critics of industrial policy claim that this is inherent to such policies, but such skepticism is unwarranted, since there are also examples of excellent strategic assessments having led to wise decisions (examples include French high-speed trains; numerical controls developed by the US Army; the Japanese semiconductor industry as guided by MITI's strategic plan) (Ergas, 1992).

Simplifying, we can say that in the 1960s, it was possible for many European countries to carry out strategic assessment based on a standard factor input-cost method. The question to be answered was: what factor inputs do we need to create so as to be combined into an industry at something close to world best practice; and how much will it cost in the national context? In the context of rapid world, and especially European, economic expansion, the main consideration for efficiency was simply to assess whether the industry could find a market enabling it to enjoy optimal scale economies, and in that context, to implement state-of-the-art production technology. Oftentimes, *filière* (commodity chain) analysis was applied to maximize the 'local content' of the target industry in the national or regional space (Salomon, 1985).

The demands placed on strategic assessment in the context of the learning economy have become vastly more complex than in the 1960s, but the techniques of assessment have not caught up. It would no longer be possible, for example, to use the same method the French employed to plan Fos-sur-Mer today, because world capacity in virtually every major sector is much closer to saturation, and there is no comfortable time lag during which policy can simply copy the best of what is being done elsewhere. The Brazilians learned this with their market protection law for computers; though it has had some considerable positive effects, it has absolutely failed to encourage competitive computer-making in that country, leaving them generations behind the state of the art (Schmitz and Cassiolato, 1992). Any strategic assessment carried out today must use the existing starting point for the economy in question, but the goal of the policy has to be to catch up somehow to a moving target, a target which will move during the period in which the policy is getting started.

The product as the central unit of reasoning

Strategic assessment has characteristically been organized around the concept of sector: can we build a computer *industry* or a shipbuilding *industry*? The advent of the learning economy means that standard sectoral-filière assessments are no longer adequate to the task. Competition via learning takes place around real products and products do not correspond necessarily to industries-filières. The majority of output of our economies is intermediate goods, and social and spatial divisions of labor create all manner of organizational clusters in the economy which do not correspond to final output sectors, or even to the grand (and now crude) distinctions between consumer and producer goods. Some of the most significant such clusters have to do with cognate intermediate products that go to very different final output sectors; they also have to do with products which have little concrete resemblance but have parallel or convergent technological trajectories, or technological complementarities.

The upshot is that the principal unit of assessment has to be the product or a technological space of products (the latter defined by spillovers, complementarities and evolutionary dynamics). This does not mean that traditional sectoral analysis is ignored. Success in a given product generally depends on the existence of a production system which extends upstream and downstream of that product in a filière, or spills over to complementary technological spaces; but this is, from an industrial policy perspective, a tactic appropriate in some cases, not a universal goal.[9]

Developmental starting points

Countries and regions have different starting points: the size of the market; the current technological, infrastructural and knowledge endowments of the society and economy; the generic image of the country or region; underlying relationships between groups and especially between organized interests; the existing stock of firms and interlinkages between them; and the nature and effectiveness of public administration. Three standard approaches to starting points can be viewed with extreme caution in light of the analysis advanced here.

The first is to reason in terms of grand categories of starting points, the principal ones of which are: big wealthy technologically endowed regions/countries; small, wealthy, technologically endowed places; big less developed or latecomer countries; small, less developed or latecomer ('less favored' in the current EU jargon) countries; and, underdeveloped/poor countries/regions. These categories have some descriptive utility, but they do not lead anywhere in particular with respect to strategies for product-based technological learning. Their principal categories – size and technology endowments – are most relevant to big, capital- and technology-intensive industries, but even there, many small, rich countries have apparently broken the size rule (Holland with Philips, Sweden with Ericsson) and many big countries have failed in spite of it (France with Thomson and Bull). They are instructive, but only up to a point.

The second, and preferable, approach is to reason in terms of broad categories of products. For products with low barriers to entry, the experiences of Italy and Germany may be guides. In the Italian cases, traditional skills were deployed in interpersonal industries, to serve a national market in the 1950s and early 1960s. That market was big and relatively fragmented (Becattini, 1975 and Nuti, 1989). Smaller countries do not have such big markets, however, and virtually all countries are more open to import competition today than was Italy in the early 1950s. The lesson is that such industries are likely to flourish only where (a) skills are good enough or highly focused enough that they can contribute something unique to the world market; (b) they can serve a local or national market which is unsatisfied by imports or can do so in a way which passes the indifference test: higher local prices are compensated by better tailoring to local demand (but with open markets and media, the knife-edge problem is sharper and sharper); or (c) where innovative institutional arrangements, such as specification subcontracting (Gereffi and Fonda, 1992), are used to link local producers to order-givers in a way that builds their skills and responsibilities.

For industries with high barriers to entry, whether because of traditional scale concerns or because of high investment in technology, the choice is a very stark one: either go all the way with a major technology policy designed to cover a technological space (for example, Airbus, the Japanese semiconductor policy, US military procurement), or target particular subsectors but still with potential for developing spillovers. It is likely that, in any country, big multinational partners will be necessary and substantial commitments of local resources over long time periods be required. The only strategies likely to succeed in the latter case are those where technological branching points (for example, which model of high-definition television? which system for transmitting mobile telephone calls?) are at hand, and where the risk is taken to develop along one branch rather than another. The optimistic note for this strategic assessment process is that there is rarely a single best world practice for any group of products.[10] Entrants can define products and practices, and they can trace out developmental pathways that continue to redefine such products and practices.

The third approach to strategic assessment is to reason in terms of norms for countries, points toward which we want to move, away from the starting point. This leads to a developmental recipe, in terms of such things as capital institutions, technological infrastructure, political and administrative institutions, entrepreneurship, and so on, which – it is said – will bring about developmental results. This is quite wrong in two respects. One is that among the successful, rich countries and regions, a great diversity of products, and hence worlds of production and accompanying economic practices and institutions, exists. They do not all follow the same rules with respect to the provision of capital, skilling of the workforce, public administration, entrepreneurship, and so on (as in Nelson, 1993). Even within given sectors, there is a plurality of successful but different models. It is a gross oversimplification, except at the most abstract level (for example, honest versus corrupt public administration; schooling versus no schooling) to try to reduce the development process to a single set of general goals with respect to different starting points. The ending points will be different, too, according to the specialization of the learning economy to be created, and the worlds of production they embody (Zysman, 1994). Those ending points are defined by assessing what kind of worlds are to be created, that is, the identities and capacities for action and coordination among the participants in the production system. A critical part of the strategic assessment of what is possible in a given time and place is, of course, talk. Technocrats may be able to offer the talking parties suggestions based on the entry conditions we mentioned above, but they cannot substitute for talk among the parties who ultimately will have to 'become' the collective actor of the world of production to be developed.

The new heterodox policy framework

In recent years, the analysis of the economic performance of certain successful industrial systems has prompted inquiry into policies and institutions that could be used to institute such systems. A new heterodox policy framework has emerged. This framework, while having many branches, shares a number of features, favoring policies that are context sensitive, that is, interested in the embeddedness of industrial practices in specific contexts and regions, hence 'bottom-up' (Thrift and Amin, 1994). It is production systems-oriented rather than firm-oriented in its focus. It has a non-Cartesian element, one that accepts the diversity of underlying technological and institutional situations of different economies. In many ways, it appears well positioned with respect to the foregoing analysis. Key words include: networks, flexibility, decentralization, cooperation, research and development, human capital, technopoles, training.

The policies are heterodox because of the kinds of public goods they would provide. In standard public goods theory, 'market failures' sometimes occur and when they do, public goods can be provided to rectify them. Such public goods must have economy-wide application, that is, they must be as *generic* as possible. The new theory also calls for policy to produce public goods, but allows that these goods may be *specific* to technological spaces: it is their developmental properties (evolution along trajectories through learning) that ultimately generalize (via spillovers and complementarities) their benefits to the wider economy and society.[11] We may now summarize the varying ingredients of this cocktail in more detail.[12]

Networking

The most widely shared element of new deal economic policies is to promote networking among firms. It is held that new forms of economic competition involve high levels of vertical disintegration and that there are extensive market failures in information exchange between firms. It follows that inter-establishment and interfirm relations and networks need to be supported to enhance their efficiency. Networking concerns both big and small firms and relations between them.

Promoting technology transfer

It is widely accepted that the rates at which technologies are absorbed by firms vary widely from place to place, especially when the economic base is composed mostly of small- and medium-sized firms. As a result, publicly funded innovation and technology transfer centers are becoming favored as means to enhance the uptake of new technologies, as well as to stimulate convergence in user–producer relations, so that incremental innovation can proceed more rapidly.

Local labor markets: training and focusing institutions

In industries with high levels of industry- or region-specific skills, but also with high levels of local labor market flexibility, there can be strong negative externalities: producers will not want to invest in adequate levels of labor training for fear of losing workers once they are trained. Moreover, in the face of rapid change in labor skills, no single employer will have the where-withal to effect the change in skill supply and lack of coordination may lead to a downward competitive spiral. Under these conditions, public institutions that provide for industry- or region-specific labor training, for strategic changes in the direction of training, and that help workers to secure jobs in the face of flexibility in specific, regionally concentrated sectors, can attenuate the effects of market failure.

Infant industry and getting a start: pre-competitive R&D and stimulating markets

Infant industries are based on new and experimental kinds of products. The probability of generating new products is high, but product configurations have not yet settled onto an identifiable technological trajectory. High levels of risk and uncertainty exist for producers in these nascent sectors. The collective effect of waiting, however, may create a vicious circle, where everyone waits for everyone else, and the overall rate of development is thereby retarded. By the same token, nations or regions that could successfully develop a new industry may find that a delayed start (especially when another region has moved ahead of them) locks them permanently out of a promising niche in the new industry. There are potential benefits to getting an early start, in contrast to this common free rider problem. Industries, firms, regions and nations that get ahead early often retain a leading position for quite some time, and in the early years there can be significant superprofits to new products. As a result, industry-specific pre-competitive R&D policies, and other policies to stimulate regional or national (often public) markets for risky new technology products, may be called for, in addition to networking and technology transfer centers.

Entrepreneurship, especially for small firms

Another element of the emerging consensus in the heterodox framework is that good ideas only become reality when potential entrepreneurs enjoy the conditions that permit them to start up firms and survive. The conditions favoring firm formation include traditional hard factors such as access to capital markets and some soft factors such as cultural images of the entrepreneur and sanctions to failure. They also include conditions such as access to information, locational sites, rules on hiring and firing labor, and access to potential customers in other firms. Entrepreneurship policies are designed to help potential entrepreneurs overcome these difficulties, although in practice the majority of them consist of loan programs for small firms.

Service centers

In the many successful Italian industrial districts, the practice of assisting existing firms in a series of concrete ways has emerged as a key method for public support of those communities of producers. Industry service centers are particularly devoted to spreading the costs for certain kinds of resources that single firms cannot afford for themselves alone, including systematic market research, foreign marketing, technology research and, in some cases, technology sharing, and on-line electronic networking facilities. In Italian regions, especially Modena, major industrial estates for small firms have been created, where state-of-the-art flexible configurations of space are made available to firms at below-market cost, not only permitting them to modernize their facilities, but also permitting them to remain together, thus enhancing communication and networking (Brusco, 1982). Service centers have also been involved in the promotion of regional brand names (something like *appellation contrôlée* for wine, but now applied to the market identities of other kinds of products), so as to enhance their non-substitutability on national and international markets.

The heterodox framework versus other policy approaches

There is hardly consensus about this sort of policy framework. Opposed to it are those whose point of departure is the rapid global transfer and diffusion of certain forms of technology and knowledge, and the increasing costs (hence entry barriers) of carrying out cutting-edge innovation projects. Their vision is of global technology-based oligopolies in competition where any national policy would have to reinforce the status of the nation's oligopolistic actors. For the USA, for example, this would mean weakening anti-trust laws (which are deemed to have outlived their purpose), either in a soft version to promote collaboration (la Sematech) or in a hard version, to promote concentration. Complementing this, some argue for neo-mercantilist trade policies. There may, in some cases, be a role for some such policies, a subject into which it is impossible to enter here. But there is little reason to believe that alone they would generate competitive technological learning and performance: the failures of concentration policy are well documented. Such failures often have to do with the absence of links between the resulting giant firms upstream to an effective national system of innovation and downstream to an effective production community. At the most, such strategies can play a limited role in the learning economy; and in some cases they may be harmful. The heterodox policy approach is quite the opposite in that it accepts as a given the openness of the trading system, the insufficiency of size alone, and in some cases, the disadvantages of concentration for competitive learning.

Another approach to policy is that of strengthening systems of innovation (SI) (Nelson, 1993). SI is also a heterodox neo-institutionalist approach to technological innovation. There are many different versions of SI extant at the present time. Almost all (except that of Edquist and Lundvall, 1993) share an emphasis on formal institutions, on scientific-engineering skills, and on the national level of formal institution-building. The major non-public institutions are firms and research laboratories; the major public ones are universities, government laboratories and procurement programs, and technical education. The heterodox approach differs in emphasis, although it is not in contradiction. The emphasis here is on: the plurality of worlds of production and innovation (science and engineering is only a part of the problem); 'small' processes of coordination via convention and relations; the circularity of conventions, relations and institutions; and hence the necessity for a significant meso-economic dimension to policy as well as a systematic national level.

The dangers of orthodoxy

The policy approach described above, like any set of measures that attempts to take a complex analysis of economic reality and create a policy formula based on it, runs the danger of missing its target. To avoid problems in the transformation of the current analysis into policy, both technical and substantive reductionism must be avoided. This analysis of learning is not inherently about small firms, networking, localism, or flexibility, per se; it is rather about adaptive technological learning in a territorial context. The proper goals of such policies are:

(a) for traditional or small-scale intermediate products, ongoing adaptation of products and processes, especially through product differentiation or moves up the price-quality curve, so as to respond to ongoing and inevitable entry by competitors, whether large firms or other regional systems; and

(b) for scale-intensive or new technology products, moves along the technological frontier, where that frontier is unknown or unknowable. The entire substantive thrust of any new-deal economic policy must be geared to these substantive goals, as specified in light of particular products and their worlds of production to be developed. New-deal economic institutions are only means to these ends.

The real danger exists, as theory now becomes packaged into policy, that such policies will become detached from this substantive content and necessary procedure of building convention, and instead devolve into mechanical formulas and self-referential content. Three particular tendencies are worrisome.

From framework to formula

Networking is now frequently discussed in policy literature as a sort of recipe for any form of communication between firms. This has no necessary relation to the precise conventional forms of interfirm relations which underlie successful worlds of production. Networking has to be a means to realizing a common developmental pathway characterized by learning. The content and shape of a network, as well as the degree of external network, will differ according to the products to be made and the specific way productive activity becomes a world of production.

The provision of *services to firms*, in substantive terms, represents a strong departure from traditional public goods provision. In the latter, public goods are provided where markets fail, due to the free rider problems that come when such goods are non-exclusive. As a result, public goods are provided which have non-specific assets: they are non-exclusive. But the provision of concrete, real services to real worlds of production involves public goods with asset specificities and, hence, a certain measure of exclusivity. It will not do to disguise service centers as mere providers of any old, generic public goods, and yet this is the tendency in the literature. Another danger is that service provision will be turned into a pretext for doing almost anything that supports local firms (especially the politically popular 'small' firms) with no substantive criteria for assessing the purpose of these services. Indeed, some services can have perverse effects. Modernization services, for example, can be used by some firms to distance themselves from local competitors and thereby to 'exit' from their local interdependencies. Many technological extension services are premised on the principle of survival for the fittest, not collective learning. Initially positive effects can then be followed by catastrophe (for example, the French *plan textile*). Services for learning in other words, only work when the goal is clear and when the services are consistent with the conventions of the world of production to be assisted, as understood by the participants themselves through talk.

Perhaps the most pernicious element of the current policy debate is the category of *small firms* as a goal. Small firms do play essential roles in many successful worlds of production. But smallness is not a goal, nor is there necessarily any commonality among firms merely because they are small. A small firm in a high-technology industry, whose products are evaluated according to their scientific-technological content, has virtually nothing to do with a small firm subcontractor to a major garment producer: their underlying conventions of work, product markets, interfirm relations, and so on, are totally different.

Finally, the category of *localism* has become important in the heterodox framework, because of its theoretical recognition of the agglomeration and territorial embeddedness or specificity of conventions in many learning-based production systems. The problem is that localism cannot be an end in itself: the territoriality of relations that go into learning is, in all cases, highly complex. If the strength of a particular local economy (one that corresponds to the jurisdictions of public institutions) is to be promoted, it is better to focus on learning and its conventional underpinnings than directly on localism. At its worst, localism could lead to an artificial closing off of the production system, reducing its flexibility and hindering the development of conventions that go beyond local borders but ultimately strengthen the local economy.

The dangers of cooperation

Nowhere is the danger greater than in the sudden stress on 'cooperation' as a key to world-class economic performance. Cooperation was, correctly, discovered as a dimension of certain kinds of highly successful industrial systems, principally in Italy and Germany, and its discovery has prompted a necessary corrective to the stress on competitive atomistic interaction of most economic thinking. But it is fully contradictory with the notions of sectoral diversity and the plurality of ways of successfully organizing modern production (the plurality of 'worlds' in other words), to hold up cooperation as a model for behavior in all cases. There are worlds of production where particular, conventionally rooted forms of cooperative coordination can be useful, but there is no general model of cooperation for all industries in all places.

Boilerplate approaches to the learning economy

The transformation of this new heterodox thinking about economic development into an all-purpose formula is dangerous as well. Porter's (1990) 'diamond' of development is perhaps the best known and certainly the most widely employed of such policy formulas. The framework has two main problems. The first is that it is superficial: for the most part, it restates the obvious *outcomes* of success as observed in many places (the four points of the diamond). It then 'reverse engineers' these outcomes, claiming that they were causes of success. But of course this does not follow. Attempting to create success by installing outcomes has always been the downfall of industrial policies; the fact that Porter has based his model on an admittedly more sophisticated analysis of real world experiences does not prevent the result from being equally prone to failure. Second, by abstracting across a wide variety of cases, that is, preferring an extensive form of reasoning to an intensive one, it turns a great diversity of experiences into a single formula, with potentially great errors (for example, all success is based on high levels of domestic intrasectoral competition, and so on). The record already shows that 'intangible factors play a central role in distinguishing cases of success in the new policy framework from dramatic failures' (Doeringer and Terkla, 1990).

Scientism and prescriptive rationality[13]

Many of the problems described above stem from the tendency of analysts to seek determinate causes of industrial performance and for policy-makers to extend that via prescriptive rationality. One of the greatest ironies observable at the present time is the transformation of an analysis explicitly inspired by 'non-Cartesian' and 'context and contingency sensitive' epistemologies into an all-purpose technocratic formula for development. The 'architecture' of a production system and of public, organized institutions is only interesting insofar as it helps understand the patterns of action which constitute production; the objects (technologies, tools, infrastructures) are themselves the outcomes of human practices and exist to facilitate such practices; rules, skills, and formal roles depend, for their efficacy on the identities of agents. The problem with much policy-oriented analysis is that it tends to reduce its vision to architecture, objects, rules, skills and roles, and to prescribe changes in them. Not only does this ignore the substantive object of analysis, but by attempting technocratically to prescribe behavior for the agents of the system, it has a high risk of failure.

The entrepreneurial state, 'Laboratories of Democracy'

The third great danger is that heterodoxy will be used as a pretext for a neo-conservative retreat of the national state from its appropriate duties, and the installation of a system of ferocious and destructive inter-regional competition. In recent years, the turn to the region in the United States was stimulated essentially by a retreat of the federal government from the economic development field. In a nation where industrial and regional policy have always been quite weak by comparison to Western Europe, this left a policy vacuum which many states and localities attempted to fill. In some cases, small tentative steps toward the support of regionally based clusters of firms have been taken, and certain elements of the policy framework described above have been set into place. But in many of the American cases, they resemble only superficially the framework discussed here; instead, what has been done is to set up public-sector centers for delivering services to individual firms on a local basis (often 'pay as you go'), where the collective and coordinative aspect of world-building has no place in the effort. These 'laboratories of democracy' (Osborne, 1992) are often nothing more than effective privatization and atomization of industrial policy (Sternberg, 1992). And recent federal programs, notably the Clinton Technology Policy, duplicate these problems by making federal resources available only via proposal-based interfirm competition (Storper, 1995).

The attempted devolution and decentralization of industrial policy in certain areas of the European Union has a different starting point, of course. Many European nations have long had more activist policy frameworks and highly centralized states. It has become imperative to allow more initiative at regional level. France, for good reason, is now more than a decade into her experiment with administrative and policy decentralization. Still the results are mixed, in part because formal devolution of decision-making does not in and of itself lead to the creation of capacities for action at the regional or local level.[14]

It is essential that such European experiments not lead to the outcome which is already observable in the United States. There, a more pernicious form of localism in economic policy is known as the 'entrepreneurial' state (Maier, 1994). The locality and its governmental agencies are viewed as collective entrepreneurs in competition with each other. In practice, this often means simply attempting competitively to lower the price of labor, taxes, and so on, so as to attract inward investment. But even when it involves such politically correct measures as stimulating firm start-ups, the ideology of interplace competition serves to block out any fruitful discussion of the need for higher levels of government to set certain ground rules (such as on interplace price competition) and to provide certain kinds of services (for example, national education policies, infrastructure, and other generic public goods). The danger exists that the European Union's principle of subsidiarity will be reduced to savage interplace price-cost competition.

This points out that subsidiarity in economic policy needs to have certain kinds of boundaries, and that what is to be built at subnational levels is not entrepreneurial states, but what we might call – taking our cue from the East Asian development literature – 'developmental states'. These distinctly activist, strategically oriented, systematic state efforts differ entirely from the notion of state entrepreneurialism and are much more consistent with the analysis developed in this paper (Wade, 1990).

Summary

At the end of this analysis, it will be helpful to draw the threads together. The institutions of the new economy consist of a complex circular relationship between specific, convention-bound, learning-oriented production systems which are themselves institutions, and various kinds of formal, organized institutions, notably firms, public governmental institutions, and other organizations such as universities, unions, and trade associations. Any policy framework which involves the creation of public institutions to build or sustain the institution of the learning economy has to be based on ways to cut into this circle, and must reject the traditional logic of 'public institution' versus 'private non-institution'.

We identified four major steps in the economic strategy in the new economy. The first is strategic assessment. The technical dimension is the determination of what kinds of products, where the product is the essential unit of analysis, and not the sector or the input–output system, are susceptible to being mastered in the economy at hand, where mastery is defined as ongoing competitive technological learning. There is a complex interaction between the product as a technology – a knowledge field – and its associated process technology. Just as products evolve through learning, so do processes, and both have dynamic parallels and complementarities which spill over their boundaries at a given moment.

However, strategic assessment is not only a technocratic task. Learning depends on the conventions which define collective identities of the actors in the production system by giving them access to a common context of coordination. Without this context, learning will fail, no matter how good the hardware is. The context cannot be produced by plans, nor bought by subsidies; in order to know whether the strategy is possible, it has to be known whether there is any reason to expect actors to go along. The circular relationship described here can only be broken into by talk. Talk is a necessary element in, and component of, strategic assessment.

The second step is the definition of the capacities for action and identities of actors which are associated with the world(s) of production to be assisted by policy. Each world – a specific, local or national concretization of Interpersonal, Market, Intellectual, or Industrial action and coordination – involves conventions, which coordinate interfirm relations, markets, labor markets, and so on. These are the substantive goals, the specific (and differentiated) end points of policy. They, too, can only be defined through the difficult and clumsy exercise of talk, in concert with analysis.

The third step is the implementation of specific versions of heterodox meso-economic policies, whose content is defined by combination of technical assessment and social process, especially talk. The substantive method of heterodox policies is not to attempt the construction of learning-based worlds of production from whole cloth, but rather to try to create precedents which build confidence and hence make possible the deepening and widening of conventions. Small experiments are one logical way to proceed.

Finally, and only at the end of this long and 'soft' process, can the need for further formal institution-building be realistically assessed and practically undertaken, the latter on the basis of confidence-precedent (and hopefully success in learning), and consequently emerging collective identities. There are other dimensions of formal institutions, that is, having to do with macro-competition rules, banking, education, and so on, which are not considered in this analysis. They, too, require links to the substantive concerns elaborated here. For example, education policies in different countries favor very different kinds of economic action, and push them down different routes of specialization. Some decisions about institutional structures at these levels can be taken with respect to strictly generic concerns (universal values of the society; inputs to any kind of modern economic activity); but a surprising number involve more concrete visions of the particular kind of productive economy and collective action which is desired. Here we have merely laid out the fragments of this way of thinking about the problem – the problem of constructing coherent conventions and frameworks of action in the learning economy.

Notes

1 The author wishes to acknowledge the support of the Institute of Geography at the University of Copenhagen, and in particular John Jorgensen, during the time this paper was prepared. Thanks are also due to the Danish Fulbright Commission, directed by Mette Skakkebaek.
2 This is, of course, the position taken by contemporary institutional economics, where uncertainty is a result of universally opportunistic behavior, generating moral hazards, which are only overcome either by authority (internalization) or the law of large numbers (Williamson, 1985).
3 This is developed in detail in Storper and Salais (1997).
4 *See* the papers in Grabher (ed.) (1993).
5 Lundvall (1990) relies in part on Habermas (1976).
6 This is the basic assumption of modern transactions-cost economics: *see above*, note 2.
7 We have a detailed study of French successes and failures in Storper and Salais (1997), Ch. 6.
8 In general, the literature treats the function of reputation as a form of cross-check on behavior. It multiplies the probability of getting caught; and multiplies the consequences of getting caught. *See* the papers in Gambetta (ed.) (1988).
9 This is meant as a deliberate critique of the 'diamond' found in Porter (1990).
10 Even Porter (1990) admits this. *See also* Zysman (1994) and Wade (1990).
11 I argue this point in greater detail in Storper (1995); a similar argument may be found in Romer (1993).
12 What follows comes largely from: Storper and Scott (1995).
13 Thrift and Amin (1994).

14 We have a detailed analysis of decentralization in Storper and Salais (1997), Ch. 11.

References

Amin, A. and N. Thrift (1993) 'Globalization, Institutional Thickness, and Local Prospects', *Revue d'Economie Régionale et Urbaine*, 3, pp. 405–30.

Arrow, K.J. (1951) *Social Choice and Individual Values*, Cambridge: Cambridge University Press.

Asanuma, B. (1989) 'Manufacturer–Supplier Relationships and the Concept of Relationship-Specific Skill', *Journal of the Japanese and International Economies*, 3, pp. 1–30.

Becattini, G. (1975) *Lo Sviluppo Economico della Toscana*, Florence: Guaraldi.

Brusco, S. (1982) 'The Emilian Model: Productive Decentralization and Social Integration', *Cambridge Journal of Economics*, 6, 167–84.

Buchanan, J. and G. Tullock (1965) *The Calculus of Consent*, Ann Arbor: University of Michigan Press.

Dei Ottati, G. (n.d.) 'Prato, 1944–1963: Reconstruction and Transformation of a Local System of Production', Florence, University of Florence, manuscript.

Doeringer, P. and D. Terkla (1990) 'How Intangible Factors Contribute to Economic Development: Lessons from a Mature Local Economy', *World Development*, 18, pp. 1295–1308.

Dosi, G., K. Pavitt and L. Soete (1990) *The Economics of Technical Change and International Trade*, New York: New York University Press.

Edquist, C. and B.-Å. Lundvall (1993) 'Comparing the Danish and Swedish Systems of Innovation', in R.R. Nelson (ed.) *National Innovation Systems*, New York: Oxford University Press, pp. 167–84.

Ergas, H. (1992) 'The Failures of Mission-Oriented Technology Policies'. Paper delivered to the international conference on 'Systems of Innovation', Bologna, October.

Ettlinger, N. (1994) 'The Localization of Development in Comparative Perspective', *Economic Geography*, vol. 70 (2), pp. 144–66.

Gambetta, D. (ed.) (1988) *Trust*, Oxford: Basil Blackwell.

Gereffi, G. (1995) 'State Politics and Industrial Upgrading in East Asia', *Revue d'Economie Industrielle*, 71, pp. 79–91.

Gereffi, G. and S. Fonda (1992) 'Regional Paths of Development', *Annual Review of Sociology*, 18, pp. 419–48.

Grabher, G. (ed.) (1993) *The Embedded Firm*, London: Routledge.

Haas-Lorenz, S. (1994) 'Apprentissage et proximité géographique dans une perspective volutionniste', doctoral thesis, University of Aix-Marseille, Faculty of Economics, Aix-en-Provence.

Habermas, J. (1976) *Connaissance et Intérêt*, Paris: Gallimard (orig. in German, 1968).

Hirschman, A.O. (1970) *Exit, Voice and Loyalty: Responses to Decline in Firms, Organizations, and States*, Cambridge, MA: Harvard University Press.

Keohane, R. (1993) 'International Institutions: Two Approaches', in J.G. Ruggie (ed.) *Multilateralism Matters: the Theory and Praxis of Institutional Form*, New York: Columbia University Press.

Krugman, P. (1990) *Rethinking International Trade*, Cambridge, MA: MIT Press.

Leamer, E.E. (1994) 'Third World Imports and the Unskilled in the West', UCLA Conference on the World Trading System after the Uruguay Round, December, UCLA Center for International Relations, Los Angeles.

Lorenz, N. (1992) 'Trust, Community, and Co-operation: towards a Theory of Industrial Districts', in M. Storper and A.J. Scott (eds) *Pathways to Industrialisation and Regional Development*, London: Routledge.

Lundvall, B-Å. (1990) 'From Technology as a Productive Factor to Innovation as an Active Process', Colloquium, Networks of Innovators, Montreal, 1–3 May.

Maier, M. (1994) 'Post Fordist City Politics', in A. Amin (ed.) *Post-Fordism: a Reader*, Oxford: Basil Blackwell.

Nelson, R. (ed.) (1993) *National Systems of Innovation*, New York: Oxford University Press.

North, D. (1981) *Structure and Change in Economic History*, New York: Norton.

Nuti, F. (1989) 'I Distretti dell'Industria Manifatturiera', Rome: report to the CNR, National Research Council of Italy.

Olson, M. (1971) *The Logic of Collective Action: Public Goods and the Theory of Groups*, Cambridge, MA: Harvard University Press.

Olson, M. (1990) 'Toward a Unified View of Economics and Other Social Sciences', in K. Alt and Shepsle (eds) *Perspectives on Positive Political Economy*, Cambridge: Cambridge University Press.

Osborne, D. (1992) *Laboratories of Democracy*, Cambridge, MA: Harvard Business School.

Patel, P. and K. Pavitt (1991) 'Large Firms in the Production of the World's Technology: an Important Case of Non-Globalization', *Journal of International Business Studies*, First Quarter, pp. 1–21.

Petit, P. (1993) 'Are Full Employment Policies *Pass?*', paper delivered to the Annual Meeting of the American Economics Association, Anaheim, CA.

Porter, M. (1990) *The Competitive Advantage of Nations*, London: Macmillan – now Palgrave Macmillan.

Putnam, R. (1992) *Making Democracy Work*, Princeton, NJ: Princeton University Press.

Rip, A. (1991) 'Meta-modeling and Technological Change', paper delivered to OECD TEP Conference, La Villette, Paris.

Romer, P. (1993) 'Implementing a National Technology Strategy with Self-Organizing Industry Investment Boards', paper prepared for the June 1993 Meeting of the Brookings Panel on Microeconomics, Washington DC.

Sabel, C. (1993) 'Constitutional Ordering in Historical Context', in F. Scharpf (ed.) *Games in Hierarchies and Networks*, Boulder: Westview Press.

Salomon, J.J. (1985) 'Le Gaulois, le Cowboy et le Samourai', report to the Ministry of Industry and Research, Paris.

Sayer, A. and R. Walker (1992) *The New Social Economy*, Oxford: Basil Blackwell.

Schmitz, H. and J. Cassiolato (eds) (1992) *High Tech for Industrial Development: Lessons from the Brazilian Experience in Electronics and Automation*, London: Routledge.

Sternberg, E. (1992) *Photoelectronics and Industrial Policy*, Albany: SUNY Press.

Storper, M. (1991) *Industrialization, Economic Development and the Regional Question in the Third World: from Import Substitution to Flexible Production*, London: Pion.

Storper, M. (1995) 'Regional Technology Coalitions: an Essential Dimension of National Technology Policy', *Research Policy*, vol. 24 (6), pp. 895–912.

Storper, M. and R. Salais (1997) *Worlds of Production: the Action Frameworks of the Economy*, Cambridge, MA: Harvard University Press.

Storper, M. and A.J. Scott (1995) 'The Wealth of Regions: Market Forces and Policy Imperatives in Local and Global Context', *Futures* (summer).

Thrift, N. and A. Amin. (1994) 'Socioeconomics, Democracy and Economic Policy', paper presented to the Innis Centennial Conference on 'Technology, Regions and Policy', Toronto, September.

Tyson, L. (1987) *Creating Advantage: Strategic Policy for National Competitiveness*, Berkeley: BRIE.

Wade, R. (1990) *Governing the Market: Economic Theory and the Role of Government in East Asian Industrialization*, Princeton: Princeton University Press.

Williamson, O. (1985) *The Economic Institutions of Capitalism*, New York: The Free Press.

Wolfe, D. (1994) 'The Institutions of the New Economy', Ottawa: Carleton University, Dept. of Political Science, manuscript.

Zysman, J. (1994) *How Institutions Create Historically-Rooted Trajectories of Growth*, Oxford: Oxford University Press.

8
The Learning Region[1]

Richard Florida

Introduction

A new age of capitalism is sweeping the globe. In Silicon Valley, a global center for new technology has emerged, where entrepreneurs and technologists from around the world backed by global venture capital invent the new technologies of software, personalized information and biotechnology that will shape our future. In the financial centers of Tokyo, New York and London, computerized financial markets provide instantaneous capital and credit to companies and entrepreneurs across the vast reaches of the world. In the film studios of Los Angeles, computer technicians work alongside actors and film directors to produce the *software* that will run on new generations of home electronics products produced by television and semiconductor companies in Japan and throughout Asia. Computer scientists and software engineers in Silicon Valley and Seattle work with computer game makers in Kyoto, Osaka and Tokyo to turn out dazzling new generations of high-technology computer games. In Italy, highly computerized factories produce designer fashion goods tailored to the needs of consumers in Milan, Paris, New York and Tokyo almost instantaneously. Teams of automotive designers in Los Angeles, Tokyo and Milan create designs for new generations of cars, while workers in Kyushu work to the rhythm of classical music in the world's most advanced automotive assembly factories to produce these cars for consumers across the globe. Throughout Japan, a new generation of knowledge workers operates the controls of mammoth automated factory complexes to produce the most basic of industrial products – steel. A new industrial revolution sweeps through Taiwan, Singapore, Korea, Malaysia, Thailand, Indonesia, and extends its reach to formerly undeveloped nations such as Mexico and China. And, once written-off industrial regions, like the former *Rustbelt* of the USA are being revived through international investment and the creative destruction of traditional industries.

Despite continued predictions of the 'end of geography', regions are becoming more important modes of economic and technological organization in this new age of global, knowledge-intensive capitalism. Although there have been numerous excellent studies of the dynamics of the individual regions, the role of regions in the new age of knowledge-based, global capitalism remains rather poorly understood. And, while several outstanding studies have chronicled the rise of knowledge-based capitalism, outlined the contours of the learning organization, and described the knowledge-creating company, virtually no one has developed a comparable theory of what such changes portend for regions and regional organization.

This chapter suggests that regions are a key element of the new age of global, knowledge-based capitalism. Its central argument is that regions are themselves becoming focal points for knowledge-creation and learning in the new age of capitalism, as they take on the characteristics of *learning regions*. Learning regions, as their name implies, function as collectors and repositories of knowledge and ideas, and provide an underlying environment or infrastructure which facilitates the flow of knowledge, ideas and learning. Learning regions are increasingly important sources of innovation and economic growth, and are vehicles for globalization. In elaborating this thesis, the following sections provide brief descriptions of the new era of knowledge-based capitalism and its global scope, before turning to our discussion of the dynamics of learning regions.

The knowledge revolution

Capitalism, as writers as diverse as Peter Drucker (1993) and Ikujiro Nonaka (1991) point out, is entering into a new age of knowledge-creation and continuous learning. This new system of knowledge-intensive capitalism is based upon a synthesis of intellectual and physical labor – a melding of innovation and production – or what I have elsewhere termed *innovation-mediated production* (Florida and Kenney, 1993). In fact, the main source of value and economic growth in knowledge-intensive capitalism is the human mind. Knowledge-intensive capitalism represents a major advance over previous systems of Taylorist scientific management or the assembly-line system of Henry Ford, where the principal source of value and productivity growth was physical labor. The shift to knowledge-based capitalism represents an epochal transition in the nature of advanced economies and societies. Ever since the transition from feudalism to capitalism, the basic source of productivity, value and economic growth has been physical labor and manual skill (*see* Lazonick, 1990; Hounshell, 1984). In the knowledge-intensive organization, intelligence and intellectual labor replaces physical labor as the fundamental source of value and profit (*see* Florida, 1991).

The new age of capitalism makes use of the entirety of human intellectual and creative capabilities. Both R&D scientists and workers on the factory floor are the sources of ideas and continuous innovation. Workers on the factory floor use their deep and intimate knowledge of machines and production processes to devise new, more efficient production processes. This new system of economic organization harnesses the knowledge and intelligence of the team – the group social mind – a sharp break with the conception of individual knowledge embodied in the lone inventor or great scientist. Teams of R&D scientists, engineers and factory workers become collective agents of innovation. The lines between the factory and the laboratory blur (Florida and Kenney, 1990; 1993).

> The factory itself is a living lab with bright capable people. The key is to use their brains. Constant improvement means constant change. You cannot get constant improvement, if you have the status quo. How do you get constant change? You get it by doing things you have never done before. Isn't that what they do in a lab? Try to figure out things they never did before (personal interview by Richard Florida, November 1990).

The knowledge-intensive factory is indeed becoming more like a laboratory – a place where new ideas and concepts are generated, tested and implemented. No longer just a place of dirty floors and smoking machines, grease, muscle and sweat, the factory is increasingly an environment of brain-power and technological innovation. A massive mobilization of workers' intelligence, not just physical skill, lies at the heart of the knowledge-intensive factory. And, this knowledge-intensive factory takes on many of the physical characteristics of the laboratory as well. Like a laboratory, the knowledge-intensive factory is an increasingly clean, technologically advanced and information-rich environment. Workers use advanced computerized equipment to monitor, control and conduct experiments on production. In fact, production itself is guided by software and computer systems which image the labor process. Workers control this advanced equipment, but seldom actually touch the work in-process. In advanced semiconductor and consumer electronics factories, workers perform their tasks in clean room environments, alongside robots and machines which conduct the physical aspects of the work. In some knowledge-intensive factories, laboratory-like spaces are available for workers, which may include sophisticated laboratory-like equipment – computerized measuring equipment, advanced monitoring devices and test equipment. Modern steel mills, for example, come equipped with technologically sophisticated ladle metallurgy facilities where factory workers are able to scientifically adjust the basic chemistry of molten steel while it is being made. Workers use these laboratory-like spaces together with R&D scientists and engineers to analyze, understand, fine-tune, and improve products and production processes.

The knowledge-intensive corporation uses new organizational structures, new incentives, and new mechanisms to elicit workers' cooperation and mobilize and harness their ideas. This goes far beyond the old feel-good techniques of labor–management cooperation, quality-of-work life, labor–management committees, and 1980s-style quality circles. This is a deep organizational refashioning which is absolutely required to make the factory a center for innovation and the constant application of intelligence. The knowledge-intensive organization also requires a new type of worker. This new environment places a premium on the ability to work effectively in teams and manipulate abstract constructs associated with digital information technology. Workers operate computerized machines, understand and program computers, and use their minds as well as their physical capabilities. These new workers require skill levels which are equivalent to the electrical engineers of two decades ago. The new worker must be trained and managed more like a university-trained researcher or engineer rather than as a traditional factory worker.

The knowledge-intensive organization uses its extensive corporate network of affiliates, suppliers, customers and related organizations as another source of knowledge and new ideas. As Kenichi Imai (1991) has shown, such networks provide a powerful dynamic for innovation, as each organization in the network faces considerable economic incentive to improve. As the R&D and technological knowledge of each organization advances, the network as a whole accumulates greater innovative capability and each organization becomes a more valuable member of the network. The network becomes a powerful arena for continuous innovation and improvement.

The advances in the organizational context for knowledge work are not just limited to the factory. They are rapidly spreading and diffusing into the organization of knowledge work in other settings including the R&D laboratory and in the office itself – the center for a tremendous amount of knowledge work. Indeed, companies such as Steelcase and Xerox are developing new ways of conceptualizing and of organizing the physical arenas and support infrastructures required for knowledge work.

Consider the traditional physical arena in which much knowledge work takes place – the office. The traditional office was and is, for lack of a better phrase, a Taylorist office – the physical manifestation of Taylorist management. Cut up into little pieces, separated by walls and partitions, it exemplified the logic of hierarchy and command and control. Work took place in isolation. Interaction was monitored and took place in formal settings. Conference rooms, for example, were the site of formal meetings where people reported information and bosses gave the orders on what to do next. Still, much of the work of the organization took place through its so-called informal structure – conversations in the hall and so on.

The new arena for knowledge work as it is being developed by companies like Steelcase and Xerox is seen as continuous interaction and knowledge-creation. The very definition of work is as dialogue and interaction. For researchers at Xerox PARC, work is defined as interacting centers of communication and communities of practices. In the words of William Miller, director of research and business development for Steelcase, work is the result of hundreds of interactions which are knowledge event driven. 'The model for knowledge work', he adds, 'is the interactive classroom' (personal interview by Richard Florida, January 1994).

The new environment for knowledge work then is evolving as an arena for dialogue. It facilitates and enhances this dialogue of knowledge – and reinforces and stores its most important content. It is a place where continuous learning occurs. Steelcase, for example, is developing new office furniture which is reconfigurable to meet the needs of rapidly changing teams – essentially an office on wheels. These reconfigurable offices provide space for individual work, thought and contemplation, and flexible group work space. Steelcase is also developing multi-media software which can be used by work groups to design their work space, and computerized tools and expert systems which enable them to measure their performance. Xerox is developing new software and media that are designed to capture and preserve the results of knowledge transfer and interaction. These include new forms of electronic white boards and active story boards, which aim to capture and preserve the rich array and diversity of exchanges among workers and work groups. The physical environment of knowledge work – the office of the future – must have physical means to store its social, group memory.

The age of continuous innovation

The rise of knowledge-intensive capitalism is giving rise to a pattern of continuous or *perpetual innovation* – an economy in which value is created through a process of rapid, continuous and accelerating technological innovation (Morris-Suzuki, 1984; 1988). Increasingly severe competition in many market segments and shorter runs make innovation an increasingly crucial source of value and competitive advantage. In digital electronics, for example, products are outmoded in a year, sometimes less. This new age of capitalism is powered by the most accelerated pace of innovation and invention in all of human history. And, this technological revolution will only accelerate with the advent of biocomputing and bioengineering which enable ever more accelerated cycles of innovation, ever more efficient production, and more complete recycling of materials and the production of *green* products (Florida, 1996; Ayres, 1989; Tibbs, 1993).

In this new environment, economic success is tied to an organization's ability to constantly improve products and processes and to rapidly deploy new products and technologies. To maintain its long-run sustainable advantage, knowledge-based organizations and firms must continuously create new products, rapidly diversify and customize existing products, open up new markets, and constantly develop and create new market opportunities. And, to do so effectively, it must have the ability to constantly improve products and processes, to revamp the production process itself, and to rapidly deploy new products and technologies.

The knowledge-based organization must be adept at developing and capturing the benefits of what Japanese innovation expert, Fumio Kodama (1991) has termed *fusion technologies*, such as mechatronics which combines microelectronics and mechanical technologies, in the form of computerized machine tools, flexible manufacturing cells, and computer-integrated factories, and optoelectronics which fuses optics and electronics. Fusion is not just a technological phenomenon; it is rapidly becoming a central dimension of industrial organization, as high-technology industries such as semiconductors, computers, software and consumer electronics blur into one another. The lines between traditional and high-technology industries are blurring as well, as automobile and steel production undergo new waves of *creative destruction* first identified by the great economist Joseph Schumpeter. In the new age of capitalism, automobile and steel companies produce software, integrated circuits, programable logic controllers, advanced robotics and machine tools and the artificial intelligence and software programs which run those various machines, tools and pieces of equipment. These companies are developing their own software capabilities, spinning-out software subsidiaries, and investing in high-technology startups as they endeavor to compete in the new age of technology-intensive, digitally based manufacturing.

The ability to turn innovations into high-quality products that the world's consumers want to buy is increasingly recognized as a critical dimension of economic performance. Innovative technology means little if it cannot be turned into high-quality manufactured products. The ability to turn innovations into products is a key to long-run sustainable advantage. The knowledge-intensive organization must therefore overcome the *breakthrough illusion* (Florida and Kenney, 1990) – the myth that new technology alone offers a source of competitive advantage. This breakthrough illusion is a major reason for the sagging competitiveness of American companies from IBM to Xerox. Silicon Valley too has fallen victim to this breakthrough illusion, as venture capitalists and high-technology start-up companies produce wave after wave of innovations, yet find it increasingly difficult to turn these innovations into successful generations of products. In these companies, entrepreneurs and R&D scientists earn huge sums of money for developing breakthroughs, while factory workers toil for low wages in dirty, sweatshop conditions. What's more, Silicon Valley's high-technology companies are plagued by an extraordinarily high rate of labor turnover – sometimes as high as 50 per cent per year – as R&D scientists and engineers move from company to company in search of higher salaries and more stock options. Such astronomical turnover can seriously disrupt ongoing R&D efforts; worse yet, it has helped to set in motion a growing wave of lawsuits over intellectual property.

The history of industrial capitalism is littered with accounts of once-powerful and innovative nations, as well as firms, which have been surpassed by less innovative but more efficient counterparts (*see* Kennedy, 1987; also Olson, 1982 on the rise and decline of nations). Britain continued for a long time to produce the world's best scientists and garner a significant share of Nobel prizes, even after it was overtaken by the United States and Germany. The United States continues to lead the world in breakthrough innovations and in basic science, but this has done little to restore competitiveness in a host of industries.

The Global Shift

Knowledge-intensive capitalism is increasingly a global system and the knowledge-intensive organization is increasingly a global actor. All aspects of R&D and product development to production and marketing must be oriented to and increasingly take place in major markets throughout the globe. The vehicle for accomplishing this is international investment, which has risen to all-time highs during the 1990s (*see* Porter, 1990; Reich, 1991; Dicken, 1992; Graham and Krugman, 1991). Indeed, according to a growing number of observers, international investment has surpassed global trade as the defining feature of the new global economy. Without question, international investment is the new and defining feature of the global economy. A recent United Nations report (1993) shows that today transnational corporations operate some 170,000 factories and branches throughout the globe. In 1992, this world-wide network of foreign affiliates generated more than $5 trillion in sales, exceeding world exports of $4 trillion, one-third of which took the form of intrafirm trade.

Globalization is increasingly taking place through the vehicle of integrated *transplant* complexes of assembly facilities and surrounding supplier and product development activities, the best examples of which include Toyota and Honda's massive production complexes in the United States. In fact, Japanese automotive production in North America is strongly concentrated in an integrated transplant complex comprising seven major automotive assembly complexes, more than 75 steel production, finishing and service center sites, and more than 400 automotive parts suppliers located in and around the traditional industrial heartland region of the US (Kenney and Florida, 1993). A similar complex is developing in and around Cardiff, Wales. Japanese semiconductor electronics and computer transplants are concentrated in a similar, though not as extensive, complex in and around Silicon Valley stretching north into Portland, Oregon and Seattle, Washington. An extensive network of Japanese electronic production extends through Southeast Asia.

Transplant investment is the source of important productivity improvement, organizational learning, and continuous improvement. According to a recent study by the McKinsey Global Institute (1993), transplants increase productivity by bringing international best-practices to foreign soil, and placing pressure on domestic industries to adopt those best-practices. The McKinsey study concludes that foreign direct investment – transplant factories – has been far more important than trade as a force for improving productivity in the advanced industrial nations, particularly the United States and Britain.

Transplants from leading-edge producers: 1) directly contribute to higher levels of domestic productivity, 2) prove that leading-edge productivity can be achieved with local inputs, 3) put competitive pressure on other domestic producers, and 4) transfer knowledge of best-practices to other domestic producers through natural movement of personnel. Moreover, foreign direct investment has provoked less political opposition than trade because it creates jobs instead of destroying them. Thus, it is likely to grow faster in years to come (McKinsey Global Institute, 1993).

A recent OECD study provides considerable evidence of the link between international investment, productivity and economic growth. Comparing investment and productivity patterns in 15 advanced industrial nations, the OECD study found that foreign-owned companies are typically more efficient than domestic firms in both absolute levels and in rates of productivity growth. The study found that these productivity gains resulted from more advanced technology than domestic industries, or from adding capacity. By contrast, productivity increases at locally owned companies more often resulted from downsizing and layoffs. The study also found that international investment has been a key source of employment growth across the advanced industrial nations. In 10 of 15 countries studied, foreign-owned companies created new employment more rapidly than did their domestically owned counterparts, sometimes expanding their operations while domestic firms were contracting. In three others, they eliminated jobs, but they did so more slowly than domestically owned enterprises. The study found that the largest employment declines occurred in Japan and Germany, where soaring costs during the 1980s caused international investors to cut a significant number of jobs. Furthermore, the OECD study points to a link between investment and trade, as foreign subsidiaries tended to export and import more than domestic firms, with most of the imports taking the form of intrafirm trade.

Over the course of the 1980s, international investment was a key factor in reversing the deindustrialization and disinvestment of US manufacturing industry. Japanese industry invested more than $25 billion in rebuilding the US automotive-related industries, making massive investments in eight major automotive assembly complexes, 400-plus automotive parts suppliers, 72 joint-venture steel facilities, and 21 rubber and tire facilities (Kenney and Florida, 1993). Japanese companies have spent more than $7 billion in a massive technological and organizational upgrading of joint-venture steel facilities. European chemical and drug companies contributed a massive amount to both the technological and production capabilities of the US drug and chemical industries. Together, Japanese and European companies helped to rebuild the television production infrastructure of the United States at the same time that US companies virtually abandoned that sector. As a result, productivity also grew more rapidly in foreign-owned manufacturing companies in the US than for the manufacturing sector as a whole during the 1980s. The real output of foreign-owned manufacturers rose nearly four times as fast as all manufacturing establishments between 1980 and 1987. International investment has generated productivity increases and value-added which outdistances US-owned companies. From 1987 to 1990, for example, the rate of increase in plant and equipment expenditures for foreign-affiliated industrial enterprises (for examples non-bank, non-agricultural business) was five times greater than that for US-owned business. As of 1989, value-added per employee was substantially higher in foreign-owned firms than for US-owned manufacturers (*see* US Department of Commerce, 1993). This is in large measure due to the transfer and use of world-class organizational and management practices.

Transplants have also played an important role in the economic resurgence of America's Industrial Midwest – a region which produced more than $350 billion in manufacturing output (making it the third largest manufacturing economy in the world) and exported more than $100 billion in goods in 1991. Once a center for traditional Taylorist manufacturing, knowledge-intensive organization is spreading rapidly through the Industrial Midwest. The findings from a survey of a significant sample of Midwestern companies indicate that more than half of the responding companies are beginning to make the transition to a knowledge-intensive organization (*see* Florida, 1995). While these companies have not completed the journey, they are clearly on a path to becoming knowledge-intensive companies. Consider the following example. In 1990, I visited a traditional American-owned automotive parts company which was beginning to supply Toyota and other Japanese transplant companies in North America. When I went back to this company in the fall of 1993, I found a true knowledge-intensive organization, supplying Japanese and Big Three car makers from multiple plants throughout the world. All of this is nothing short of amazing, in a region that less than five years ago was virtually completely locked into traditional Taylorist management styles. International investment is a key competitive advantage of this revitalized Industrial Midwest region. It may well be the world's only major manufacturing center which has a sizeable concentration of the world's best companies – from the US, Japan and Europe – in one region.

To achieve sustainable advantage in an age of knowledge-intensive capitalism, firms and organizations are also moving to globalize R&D activities, by establishing R&D facilities in the leading technological centers of the world. Such globalization of R&D is required to tap into the sources of knowledge and ideas, and scientific and technical talent which are embedded in cutting-edge regional innovation complexes such as Silicon Valley in the United States, Tokyo or Osaka in Japan, Stuttgart in Germany, and many others. That is why IBM conducts a considerable amount of its personal computer R&D in Japan, and why more and more US and European companies are establishing R&D branches in Japan. In fact, over the past five years, more and more Japanese transplants have opened R&D facilities in the United States. There are some 250 stand-alone R&D facilities in the United States (*see* Florida, 1997). Leading European chemical and pharmaceutical companies also operate another 100 or so R&D facilities in the US, accounting for roughly half of all US R&D spending in these sectors. For the most part, international corporations have chosen to locate their R&D facilities in or near major US technology centers, with electronics R&D located in the high-technology complex of Silicon Valley, automotive body design located outside of Los Angeles, automotive product development located in the automotive technological complex around Detroit, and the chemical and pharmaceutical industry in the New Jersey area.

Toward the learning region

The shift to knowledge-intensive economic organization goes beyond the particular business and management strategies of individual firms. It must involve the development of new inputs and a broader infrastructure at the regional level on which individual firms and production complexes of firms can draw. The nature of this economic transformation makes regions key economic units in the global economy (*see* Wolfe, 1994; Cooke, 1994; Federal Reserve Bank of Chicago, 1993). In essence, globalism and regionalism are part of the same process of economic transformation. In an important and provocative essay in *Foreign Affairs*, Kenichi Ohmae (1993), the former director of McKinsey's Tokyo office suggests that regions or what he calls *region-states* are coming to replace the nation state as the centerpiece of economic activity.

The nation state has become an unnatural, even dysfunctional unit for organizing human activity and managing economic endeavor in a borderless world. It represents no genuine, shared community of economic interests; it defines no meaningful flows of economic activity. On the global economic map the lines that now matter are those defining what may be called region states. Region states are natural economic zones. They may or may not fall within the geographic limits of a particular nation – whether they do is an accident of history. Sometimes these distinct economic units are formed by parts of states. At other times, they may be formed by economic patterns that overlap existing national boundaries, such as those between San Diego and Tiajuana. In today's borderless world, these are natural economic zones and what matters is that each possesses, in one or another combination, the key ingredients for successful participation in the global economy.

Region states, Ohmae points out, are fundamentally tied to the global economy through mechanisms such as trade, export, and both inward and outward foreign investment – the most competitive region states are home not only to domestic or indigenous companies, but are attractive to the best companies from around the world. Region states can be distinguished by the level and extent of their insertion in the international economy and by their willingness to participate in global trade.

The primary linkages of region states tend to be with the global economy, and not with host nations. Region states make such effective points of entry into the global economy because the very characteristics that define them are shaped by the demands of that economy. Region states tend to have between five million and 20 million people. A region state must be small enough for its citizens to share certain economic and consumer interests but of adequate size to justify the infrastructure – communications and transportation links and quality professional services – necessary to participate economically on a global scale. It must, for example, have at least one international airport and, more than likely, one good harbor with international-class freight-handling facilities. A region state must also be large enough to provide an attractive market for the broad development of leading consumer products. In other words, region states are not defined by their economies of scale in production (which, after all, can be leveraged from a base of any size through exports to the rest of the world) but rather by having reached efficient economies of scale in their consumption, infrastructure and professional services.

The new age of capitalism requires a new kind of region – *learning* or *knowledge-creating* regions. In fact, regionally based complexes of innovation and production are increasingly the preferred vehicle used to harness the knowledge and intelligence embedded in learning regions across the globe. For most of this century, successful regional as well as national economies grew by extracting natural resources such as coal and iron ore, making materials such as steel and chemicals, and manufacturing durable goods such as autos, appliances and industrial machinery. The wealth of regions and of nations in turn stemmed from their abilities to leverage so-called natural comparative advantages that allowed them to be a mass producer of commodities competing largely on the basis of relatively low production costs. Regional development policies emphasized financial incentives, reduction of the marginal costs of operation, regulation of business externalities, and recruitment of branch plant firms.

Over the past two decades, as we have seen, the nature of economic progress has been redefined and has changed fundamentally. Global competition has shifted the base of manufacturing activity, as foreign competitors have succeeded in making lower-cost commodities to challenge the region's companies in both domestic and international markets. Global markets demand increasingly sophisticated, high-quality, high value-added products and services, not just mass-produced commodities. In this new environment, the nexus of sustainable economic advantage rests upon the ability to create and harness knowledge in new product design and development, and continuous improvement of products and processes on the factory floor.

Regions in a nutshell serve as collectors of knowledge, intellectual capacity, and creative capabilities.

To be effective in this increasingly borderless global economy, regions must be defined by the same criteria and elements which constitute a knowledge-intensive firm: continuous improvement, new ideas, knowledge-creation and organizational learning.

Regions must adopt the principles of knowledge creation and continuous learning, they must in effect become *knowledge-creating* or *learning regions*. As such, regions must provide a series of related infrastructures which can facilitate the flow of knowledge, ideas and learning. The role of economic development policy must be to develop an institutional framework that can facilitate the shift to this new production system, and, just as important, to eliminate any remaining public and regulatory barriers which inhibit its emergence.

All regions possess a basic set of ingredients that constitute a production system (Table 8.1). They all have a *manufacturing infrastructure* – a network of firms that produce goods and services. Mass-production organization was defined by a high degree of vertical integration and internalization of capabilities. External supplies tended to involve ancillary or non-essential elements, were generally purchased largely on price, and stored in huge inventories in the plant. Knowledge-intensive economic organization is characterized by a much higher degree of reliance on outside suppliers and the development of co-dependent complexes of end-users and suppliers. In heavy industries, such as automobile manufacturing, large assembly facilities play the role of hub, surrounding themselves with a spoke network of customers and suppliers in order to harness innovative capabilities of the complex, enhance quality and continuously reduce costs.

Regions have a *human infrastructure* – a labor market from which firms draw knowledge workers. Mass production industrial organization was characterized by a schism between physical and intellectual labor – a large mass of relatively unskilled workers who could perform physical tasks but had little formal involvement in more managerial, technical or intellectual activities, and a relatively small group of managers and executives responsible for planning and technological development. The human infrastructure system of mass production – the system of public schools, vocational training, and college and university professional programs in business and engineering – evolved over time to meet the needs of this mass-production system turning out a large mass of cogs-in-the-machine and a smaller technocratic elite of engineers and managers. The human infrastructure required for a learning region is quite different. As its name implies, a learning region requires a human infrastructure of knowledge workers who can apply their intelligence in production. The education and training system must be a learning system that can facilitate life-long learning and provide the high levels of group-

Table 8.1 From Mass Production to Learning Regions

	Mass-production region	Learning/knowledge-creating region
Basis of competitiveness	Comparative advantage based upon: * natural resources * physical labor	Sustainable advantage based upon: * knowledge creation * continuous improvement
Production system	Mass production * physical labor as source of value * separation of innovation and production	Knowledge-based production * continuous creation * knowledge as source of value * synthesis of innovation and production
Manufacturing infrastructure	Arm's length supplier relations	Supplier systems as a source of innovation
Human infrastructure	* low-skill low-cost labor * Taylorist work force * Taylorist education and training	* knowledge workers * continuous improvement of human resources * continuous education and training
Physical and communication infrastructure	Domestically oriented physical infrastructure	Globally oriented physical and communication infrastructure
Industrial governance system	* adversarial relationships * top-down control	* mutually dependent relationships * network organization
Policy system	Specific retail policies	Systems/infrastructure orientation

orientation and teaming required for knowledge-intensive economic organization.

All regions possess a *physical and communications infrastructure* upon which organizations deliver their goods and services and communicate with one another. The physical infrastructure of the mass production facilitated the flow of raw materials to factory complexes and the movement of goods and services to largely domestic markets. Knowledge-intensive firms are global players. Thus, the physical infrastructure of the new economy must develop links to and facilitate the movement of people, information, goods and services on a global basis. Furthermore, a knowledge-intensive organization draws a great portion of its power from the rapid and constant sharing of information and increasingly electronic exchange of key data between customers, end-users and their suppliers. For example, Johnson Control's factory which manufactures seats for Toyota's Kentucky assembly plant receives a computerized order for seats as each new Camry begins its way down the assembly line. A learning region requires a physical and communication infrastructure which facilitates the movement of goods, people and information on a just-in-time basis.

To ensure the growth of existing firms and birth of new ones, all regions have a capital allocation system and financial market. One of the existing weaknesses in the US is that financial systems are creating impediments to the adoption of new management practices. For example, our interviews with executives and surveys of knowledge-intensive firms indicate that banks often require inventory to be held as collateral, creating a sizeable barrier to the just-in-time inventory and supply practices which define knowledge-intensive economic organization.

All regions provide mechanisms for *industrial governance* – formal modes and informal patterns of behavior between and among firms, and between firms and government organizations. Mass-production regions were characterized by top-down relationships, vertical hierarchy, high degrees of functional or task specialization, and command-and-control methods of organizing. Learning regions must develop governance structures which reflect and mimic those of knowledge-intensive firms, that is, co-dependent relations, network organization, flat or lean organization, and a focus on customer requirements. This goes for government and non-profit organizations, particularly economic development organizations, as well as for private enterprises.

Learning regions provide the crucial inputs required for knowledge-intensive economic organizations to flourish: a manufacturing infrastructure of interconnected vendors and suppliers; a human infrastructure that can produce knowledge workers, facilitate the development of a team-orientation, and which is organized around life-long learning; a physical infrastructure which facilitates and supports constant sharing of information, electronic exchange of data and information, just-in-time delivery of goods and services, and integration into the global economy; and a capital allocation and industrial governance system attuned to the needs of knowledge-intensive organizations.

Building the future

For most of the past two decades, scholars predicted a shift from manufacturing to a post-industrial service economy, or from basic industries to high technology. In the wake of the predictions, efforts were undertaken to invest in new critical technologies and industries. But, the change underway is not one of old sectors giving way to new, but a more fundamental change in the way goods are produced and the economy itself is organized – from a mass production to a new knowledge-based economy. The implications of the epochal economic transformation are indeed sweeping.

With the new century, organizations will accelerate their shift toward the principles of knowledge-based organization, utilizing management systems which harness knowledge and intelligence at all points of the organization from the R&D laboratory to the factory floor. Being lean or efficient is not enough. A knowledge-intensive organization must be a source of constant learning and a continuous stream of new knowledge and ideas. The ability to continuously reorganize and restructure to meet the needs of new markets and new customer demands is another key characteristic of the knowledge-intensive company. The ability to bolster short-run profits or attain short-run competitive advantage will no longer do.

At a broader level, we are likely to see a shift from strategies and policies which emphasize competitiveness to ones which revolve around the concept of *sustainable advantage*. Sustainable advantage means that organizations, regions and nations shift their focus from short-run economic performance to recreating, maintaining and sustaining the conditions required to be a world-class performer through continuous improvement of technology, continuous development of human resources, the use of clean production technology, elimination of waste, and a commitment to continuous environmental improvement. Indeed, the concepts of sustainable advantage and of sustainable development have the potential to become central organizing principles for economic and political governance at the international, national and regional scales. In this sense, there is some possibility that over time they may come to replace the increasingly dysfunctional *Fordist* model of political-economic regulation.

Knowledge-intensive organizations will face many challenges in the twenty-first century. Maintaining a balance between cutting-edge innovation and high-quality and efficient production may be the most critical. To do so may require a melding of best-practice techniques throughout the world – the creation of new hybrid forms of knowledge-intensive organizations. Such organizational mechanisms would blend the ability of leading-edge US high-technology companies to spur individual genius and creativity, with strategies and techniques for continuous improvement and the collective mobilization of knowledge.

Developing an effective approach to globalization is another major challenge facing knowledge-intensive organizations. Knowledge-intensive firms and organizations will build increasingly dense global webs of R&D and manufacturing capabilities. Regions will increasingly become complexes of international companies tied through international investment, trade and alliances to other global regions. Spanning the globe in its quest for new sources of knowledge and ideas, knowledge-intensive organizations must work to create truly knowledge-creating regions – learning regions – which can supply the requisite human, manufacturing and technological infrastructures required to support knowledge-intensive production.

The nature of government is also likely to change. The nation-state is being squeezed between the poles of the need for expanding international governance mechanisms and the increasing role of regions in the global economy.

For the past century, government economic policy has revolved around the principles of an older, bygone stage of capitalism. Like any Taylorist organization, it has been organized around the principles of centralization, command-and-control, and functional specialization. Such policy will have to change to meet the needs and demands of the new economy. That means government must concentrate on building the broad infrastructure of technological, manufacturing and human capabilities required for knowledge-intensive capitalism. Ultimately, government itself will have to become a network of knowledge-intensive organizations. The ability for government to adopt the principles of continuous knowledge mobilization and knowledge-intensive organization will become a source of sustainable advantage for firms, regions and nations in the twenty-first century.

But without doubt, the greatest challenge of the early twenty-first century will be the creation of new and more effective mechanisms for international cooperation in economic and political affairs. Twenty-first century society will require a new global organizational and institutional framework to orient and structure trade, investment, environmental and security considerations. The structural weaknesses of the current system are increasingly an obstacle to the further evolution of knowledge-based capitalism.

The social and economic systems of the twenty-first century will be remarkably different from those which have operated for most of the twentieth century. Knowledge and human intelligence will replace physical labor as the main source of value. Technological change will accelerate at a pace heretofore unknown – innovation will be perpetual and continuous. Knowledge-intensive organizations based upon networks and teams will replace vertical bureaucracy, the cornerstone of the twentieth century. The intersection of relentless globalization and the densely concentrated transplant production complexes will gradually erode the power and authority of the nation state – the paragon of nineteenth and twentieth century political economy. Whole new institutions for international trade, investment, environment and security will doubtless be created. While the new century holds out great hope, it will require tremendous energy and effort to set in motion the necessary changes, and an unparalleled collective effort to bring them about.

Note

1   This is a revised version of an article that originally appeared in *Futures* (June 1995), 27, 5, pp. 27–36.

References

Ayres, R. (1989) 'Industrial Metabolism', in J.M. Ausubel and H.E. Sladovich (eds) *Technology and Environment*, Washington DC: National Academy Press.

Bell, D. (1973) *The Coming of Post-Industrial Society: a Venture in Social Forecasting*, New York: Basic Books.

Block, F. (1990) *Postindustrial Possibilities: a Critique of Economic Discourse*, Berkeley, CA: University of California Press.

Bloomquist, L. *et al.* (eds) (1992) *Productivity in Knowledge-Intensive Organizations: Integrating the Physical, Social and Informational Environments*, Ann Arbor, MI: Industrial Technology Institute.

Bluestone, B. and B. Harrison (1982) *The Deindustrialization of America: Plant Closings, Community Abandonment, and the Dismantling of Basic Industry*, New York: Basic Books.

Cooke, P. (1994) 'The New Wave of Regional Innovation Networks', Unpublished Working Paper, Centre for Advanced Studies, University of Wales-Cardiff, March.

Dertouzos, M.L., R. Lester and R. Solow (1989) *Made in America: Regaining the Productive Edge*, Cambridge, MA: MIT Press.

Dicken, P. (1992) *Global Shift: industrial change in a turbulent world*, London: Guilford Press.

Drucker, P.F. (1993) *Post Capitalist Society*, New York: Harper Business.

Federal Reserve Bank of Chicago (1993) *Regional Economies in Global Markets*, Chicago: Federal Reserve Bank of Chicago.

Florida, R. (1991) 'The New Industrial Revolution', *Futures*, vol. 23 (6), July–August, pp. 559–76.

—— (1994) *International Investment: Neglected Engine of the Global Economy*, Washington DC: American Enterprise Institute.

—— (1995) 'The Industrial Transformation of the Great Lakes Region', in P. Cooke (ed.) *The Rise of the Rustbelt*, London: UCL Press, pp. 162–76.

—— (1996) 'Regional Creative Destruction: Production Organization, Globalization, and the Economic Transformation of the Industrial Midwest', *Economic Geography*, vol. 72 (3), July, pp. 314–34.

—— (1996) 'Lean and Green: the Move to Environmentally-Conscious Manufacturing', *California Management Review*, vol. 39 (1), Fall, pp. 80–105.

—— (1997) 'The Globalization of R&D: Results of a Survey of Foreign-Affiliated R&D Laboratories in the USA', *Research Policy*, vol. 26 (1), March, pp. 85–103.

Florida, R. and M. Kenney (1990) *The Breakthrough Illusion: corporate America's failure to move from innovation to mass production*, New York: Basic Books.

—— (1991) 'Transplanted Organizations: the Transfer of Japanese Industrial Organization to the United States', *American Sociological Review*, vol. 56(3) (June), pp. 381–98.

—— (1993) 'The New Age of Capitalism: Innovation-Mediated Production', *Futures*, vol. 25(6), (July–August) pp. 637–51.

—— (1994) 'The Globalization of Japanese Research and Development – the Economic Geography of Japanese Investment in the United States', *Economic Geography*, vol. 70 (4), October, pp. 344–69.

Graham, E.M. and P. Krugman (1991) *Foreign Direct Investment in the United States*, Washington DC: Institute for International Economics.

Hounshell, D. (1984) *From the American System to Mass Production, 1800–1932: the Development of Manufacturing Technology in the United States*, Baltimore: Johns Hopkins University Press.

Imai, K. (1991) 'Globalization and Cross-Border Networks of Japanese Firms', prepared for the conference on 'Japan in the Global Economy', Stockholm School of Economics, 5–6 September.

Keatley, R. (1994) 'OECD Says Foreign Investment is Good for You', *Dow Jones/Wall Street Journal Electronic News Service*, 8 April.

Kennedy, P. (1987) *The Rise and Fall of Great Powers: Economic Change and Military Conflict from 1500 to 2000*, New York: Random House.

Kenney, M. and R. Florida (1993) *Beyond Mass Production: the Japanese System and Its Transfer to the United States*, New York: Oxford University Press.

Kenney, M and R. Florida (1994) 'The Organization and Geography of Japanese R&D: Results from a Survey of Japanese Electronics and Biotechnology Firms', *Research Policy*, vol. 23 (3), May, pp. 305–23.

Kodama, F. (1991) *Analyzing Japanese High Technologies: the Techno-Paradigm Shift*, London: Pinter.

Lazonick, W. (1990) *Competitive Advantage on the Shopfloor*, Cambridge: Harvard University Press.

McKinsey Global Institute (1993) *Manufacturing Productivity*, Washington, DC: McKinsey Global Institute, October.

Morris-Suzuki, T. (1984) 'Robots and Capitalism', *New Left Review*, vol. 147, pp. 109–21.

—— (1988) *Beyond Computopia: Information, Automation and Democracy in Japan*, London: Kegan Paul International.

Nonaka, I. (1991) 'The Knowledge Creating Company', *Harvard Business Review*, November–December, pp. 69–104.

Ohmae, K. (1993) 'The Rise of the Region State', *Foreign Affairs*, vol. 72 (2), Spring, pp. 78–87.

Olson, M. (1982) *The Rise and Decline of Nations: Economic Growth, Stagflation, and Social Rigidities*, New Haven: Yale University Press.

Piore, M. and C. Sabel (1984) *The Second Industrial Divide: possibilities for prosperity* (New York: Basic Books).

Porter, M. (1990) *The Competitive Advantage of Nations*, New York: Free Press.

Reich, R. (1991) *The Work of Nations: Preparing Ourselves for the 21st Century Capitalism*, New York: Knopf.

Sayer, R.A. and R. Walker (1992) *The New Social Economy: reworking the division of labor*, Oxford: Basil Blackwell.

Schumpeter, J. (1942) *Capitalism, Socialism and Democracy*, New York: Harper and Row.

Storper, M. and R. Walker (1989) *The Capitalist Imperative: Territory, Technology, and Industrial Growth*, Oxford: Basil Blackwell.

Sutherland, D. (1992) 'Technology and the White-Collar Paradox: Time, Tools and the Mind's Best Work', in L. Bloomquist *et al.* (eds) *Productivity in Knowledge Intensive Organizations: Integrating the Physical, Social and Informational Environments*, Ann Arbor, Michigan: Industrial Technology Institute.

Tibbs, H. (1993) 'Industrial Ecology', (Global Business Network).

United Nations (1993) *World Investment Report, 1993: Transnational Corporations and Integrated International Production*, New York: United Nations.

United States Department of Commerce (1993) *Foreign Direct Investment in the United States*, Washington DC: US Government Printing Office, June.

Wolfe, D. (1994) 'The Wealth of Regions', paper presented at the *Workshop on Institutions of the New Economy*, Canadian Institute for Advanced Research, Toronto, 21–23 May.

Womack, J.P., D.T. Jones and D. Roos (1990) *The Machine That Changed the World: Based on the Massachussetts Institute of Technology 5-million Dollar 5-year Study on the Future of the Automobile*, New York: Rawson Associates.

Zuboff, S. (1989) *In the Age of the Smart Machine: the Future of Work and Power*, New York: Basic Books.

9
Regional Innovation Systems and Regional Competitiveness

Philip Cooke

Introduction

It is well known that the European Union is anxious to close the competitiveness gap with Japan and the USA (EC, 1994). A major part in the disparity between European competitiveness and that elsewhere concerns perceived innovation deficits (EC, 1995). These two are considered also to be of fundamental importance to the improvement of social cohesion since, it is presumed, better economic performance is associated with improved prospects for employment (EC, 1996). The question of what kinds of innovation improve prospects for both competitiveness and cohesion is an important one (Edquist, 1997; MacIntosh and Francis, 1997). A problem in the past, however, has been that growth, competitiveness and employment gains have tended to be geographically overconcentrated. This leads to the kinds of agglomeration diseconomies suffered by cities like Tokyo, prompting attempts to decentralize growth opportunities to other regions. But when this has been stimulated by attempts to decentralize science and technology infrastructures, as in Japan and France, the results have often been disappointing.

This is now understood as a failure of top-down, linear thinking about economic development processes. Increasingly, in a world of interactive learning and innovation, firms are seeking to become more 'embedded' in a regional milieu where they can build up close supplier, training and innovation links. This is because, more and more, they seek to externalize production. Where this occurs, the regional economy where they locate becomes more specialized in not only production but also the soft infrastructural support aspects of business activity. As Krugman (1995) presents it, increases in intra-industry trade produce increasing returns to scale in a world of imperfect knowledge, and endogenous technical change reinforces a tendency to spatial monopoly. But the new kind of monopoly is more specialized than the universal monopoly that many large metropolitan centres used to enjoy. So regions have the capability to become specialist spatial monopolies. Global free-trade and the growing efforts of regional agencies and governments to promote the competitive advantage of 'their' region further reinforce these tendencies with respect to the attraction of Foreign Direct Investment (FDI).

If this theory is true (and we have to accept that it may not be) then its implications for regions are profound. From being a *tabula rasa* on which are inscribed the results of past resource-based business decisions, decentralization effects of central government decisions, and the decisions of both indigenous smaller firms and indigenous or FDI large firms, the region now becomes a proactive space in which all of its assets are mobilized to try to secure regional economic competitiveness. Thus the building up of a strong cultural offer, the integration of universities with industrial requirements, and the focused training of young people and older unemployed people to fit into the new occupational needs of firms becomes more pronounced. Competitiveness as a regional attribute becomes a product of systemic interaction between diverse players who must be 'associative', 'networking', and consensus-minded. At the heart of this is the desire, if not the imperative, to be seen as innovative and supportive of innovation by firms and other organizations. To what extent can regions really achieve this?

In this chapter, the results of a large-scale EU-funded research project on 'Regional Innovation Systems' will be summarized and an attempt made to judge the degree to which diverse European regions match up to the theory and practice of 'the new regionalism'. The research examined nine EU regions and two from Central and Eastern Europe and sought to find out the extent to which the competitiveness of regions was related to their degree of systemic innovation capability. Some surprising results ensued from the studies, both in terms of the competitive posture of European firms and their culture of innovation. Certain of these results suggest reasons for a relatively weak innovative capacity among European firms. Among the stronger findings was the conclusion that where regional governances are weak or passive, associativeness tends to be low. Yet the capability of firms to solve innovation problems internally is also low while their willingness to engage in cooperative solutions is high, but frustrated. The chapter proceeds by examining competitiveness problems of regional firms in the study, followed by an exploration of innovation problems. The conclusion is preceded by a section on policy-issues for regions.

Theoretical constructs of the new regionalism

The argument here links five fundamental theoretical propositions into a construct that aims to explain the role of proximity in economic competitiveness and the particular role of regions as organizational mediators. In a period of heightened, albeit asymmetrical, globalization in which innovation and supply-chain management are key factors of competitive advantage, the region takes on an extra salience in economic coordination.

Agglomeration economies

Since Marshall (1919), spatial agglomeration has been understood as a resolution of the economic problem of minimizing relational costs between agents engaged in economic transactions. Following a lengthy period of relative underdevelopment and lack of theoretical progress, the work of Krugman (1995) has revitalized the field. As a 'new neoclassical' promulgating 'new growth theory', Krugman is not without his critics, who point to a certain sleight-of-hand in his incorporation of hitherto heretical notions like imperfect knowledge and increasing returns to scale into neoclassical growth models. Formerly, such models eschewed disequilibrium but apparently now can be made to absorb concepts like 'cumulative causation' which is a profoundly non-equilibrium idea. Nevertheless, Krugman's work is interesting for the following reasons: it links regional economic theory to theories of international trade, it recognizes the importance of specialization or 'stickiness' in economic development, and it produces an, albeit abstract, conception of regional economy which is superior to that produced in neoclassical growth theory of the old school.

So what does Krugman tell us? First, recognizing disequilibrium as a pervasive spatial condition, he stresses the importance of externalities to the cumulative causation process that creates it. He links this to his solution to the problem of neoclassical economic theory which had failed to explain intra-industry trade, that is, the fact that most trading occurs between economies trading similar not different goods and services. The solution lies in imperfect competition and increasing returns to scale, in other words the drive towards *monopoly* over specialized markets. Hence agglomerations – the most obvious instances of which are cities – arise from the development of monopoly, based on specialization with associated externality-effects and increasing returns to scale. Intra-industry trade is the key means by which disadvantaged locations overcome their disparities.

Further elaborations upon this basic model are supplied by Malmberg and Maskell (1997). They test out the Krugman thesis on specialization and find it supported for Scandinavian countries over the past 25-year period. That is, economic activity has become less diffuse and more specialized in particular localities over time. However, these emergent local monopoly conditions also give rise to barriers to communication even within, let alone between, different agglomerations. They refer to 'asset stock accumulation' which makes imitation of new knowledge or tacit knowledge difficult or impossible, other than to those with knowledge or information of equivalent value to exchange. Thus transaction costs in agglomerations may not be as low as they could be because information does not formally flow around the system like a free-good. Whether 'leakage' occurs informally is a question not addressed by the authors, but it is difficult to see how it does not. The key point to bear in mind analytically is that regions can act as spatial monopolies over certain kinds of goods and services and they reinforce this characteristic through exploitation of the externality-effects of spatial agglomeration. In Krugman's terms, development occurs in other agglomerations through intra-industry trade which broadens the market but tends to maintain, for lengthy periods of time, initial comparative advantage.

Interaction effects

Innovation, business creativity and relational contracting services all benefit in certain ways from geographical proximity. In particular, they benefit from the capability to exchange tacit knowledge relatively easily and flexibly where complementary activities are spatially co-located. Tacit knowledge is distinct from codified knowledge. The latter is found in a manual, directory or piece of software code that has been structured, analysed, and put into written format. Tacit knowledge is the relatively unformulated, interiorized, even unsuspected know-how that becomes formulated in conversations, in responses to questions and in processes of speculating, hypothesizing, or brainstorming. Disclosure of tacit knowledge demands conditions of spatial proximity among interlocutors of consequence to the problem or project at hand.

For this kind of exchange to occur comfortably, those in the interactive setting must observe certain conventions, perhaps regarding confidentiality, intellectual property, disclosure or exploitation. These may even have to be more contractually embedded as 'rules of the game' in sensitive situations. But, as it has been put in an analysis of the ways in which such conventions enable transaction cost gains to be made through the involvement of actors in 'untraded interdependencies': 'The greater the substantive complexity, irregularity, uncertainty, unpredictability and uncodifiability of transactions, the greater their sensitivity to geographical distance' (Storper and Scott 1995).

In other words, codified knowledge represents greater simplification, regularity and predictability and can thus be accessed, comprehended and used at virtually any distance, while tacit knowledge is extremely difficult to recognize, let alone communicate, other than in face-to-face meetings.

Institutional learning

Inside the firm, a great deal of informal learning proceeds all the time. In 'learning organizations' this is codified into sets of procedures whereby employees and/or processes may be monitored or new workers may be mentored. From the resulting evaluative processes, improvements may be made. However, firms are hierarchical organizations and authority flows top-down, ensuring that such assurance practices are conducted and certified. Externalized learning of the kind we are primarily concerned with is a more unstructured process, though certain kinds of organizations may animate or facilitate learning processes. Governance organizations also monitor the relative performance in specific functional areas of the cities for which they are responsible. Associations of creative artists or professional scientists and technologists facilitate knowledge diffusion in both formal and informal ways. Thus benchmarking, both internally (for example, what does this region do exceptionally well and can that be emulated in other areas?) and externally (for example, what innovations have other regions successfully implemented and what lessons can this city learn?), is of fundamental importance to what Johnson (1992) calls 'institutional learning'. This is a rather more nuanced concept than the more familiar and ultimately questionable practice of 'institutional borrowing'.

Associative governance

Accomplished regions have a dense network of governance bodies ranging from legitimate and formal authorities with a democratic mandate, such as the regional administration, to no less formal but specialist bodies or associations that are responsible for the representation of specific interests. One of the distinctive features of state-planned economies was their absence of regional public or private self-governing associations. A feature of mature and accomplished liberal democracies is the presence of a rich democratic fabric composed of associative governance bodies. In his study of economic development and democracy in Italy, Putnam (1993) noted how the most accomplished parts of the country were those which combined economic prosperity with intense civic engagement. Civic culture involves associative governance and self-regulation by responsible groups in society. This allows for the exploitation of what Putnam calls 'social capital', the latent social energy for promoting and conducting needed or desired actions or projects through social engagement involving norms of trust, reciprocity and reliability.

Given that culture can easily be conceived analytically as a separate social sphere from those of the economy and politics, it is not surprising that it is often the most replete with associational activity of many kinds. Civic engagement around regional and urban cultural activities and even more so, the promotion of policies to strengthen the regional industries is thus a common feature of advanced metropolitan regions. Science and technology, too, is represented by professional associations and promoted to government and industry in terms of future policy requirements. And like all industrial sectors, financial and producer services industries have their representative organizations and associations. The evolution of developmental capability is thus heavily contingent upon the emergence of associative governance with its capacity for realizing the surplus embedded in social capital. Cities, and regions, with agglomerative, interactive and institutional learning attributes are of the first importance to the development of associative governance competences.

Proximity capital

In a recent study of the reasons why certain kinds of agglomerations, in which small and medium-sized enterprises predominate, perform better than others, Crevoisier (1997) developed the concept of 'proximity capital'. That is, he found that for such firms access to investment capital is always one of the key problems hindering their future growth capacity. Barriers are normally placed in their path by banks that appear risk-averse or require conditions to be met that such firms find onerous. However, where an agglomeration has proximity capital, this problem is often removed. It comprises investment funds deriving from persons with their own capital to invest (such as 'business angels') and companies or organizations that maintain long-term relationships with firms not only on financial, but also broader managerial, matters. These relations may well be cooperative, community-based in origin, and may remain relatively informal or become more formalized as partnerships. The entrepreneur uses force of personality and persuasiveness to convince the investor to invest.

This is paralleled to a great extent with respect to the intellectual, cultural, and financial capital that underpin regional innovation and development. Proximity, in combination with the elements of professional community or associationism by which such activities are characterized, makes for a potential 'synergetic surplus' arising from the capabilities of symbolic actors in science and industry to access not only the physical but also the symbolic capital that may be latent until activated.

Conditions and criteria for regional innovation systems

In considering the prospects for regional systems of innovation, theoretical exploration has been performed of the key organizational and institutional dimensions providing for strong and weak regional innovation systems potential. This is a preliminary attempt to specify desirable criteria upon which systemic innovation at the regional level may occur (Cooke *et al.*, 1998). These can be divided into infrastructural and superstructural characteristics.

Infrastructural issues

The first infrastructural issue concerns the degree to which there is regional financial competence. This includes private and public finance. Where there is a regional stock-exchange, firms, especially SMEs, may find opportunity in a local capital market. Where regional governments have jurisdiction and competence, a regional credit-based system in which the regional administration can be involved in co-financing or provision of loan-guarantees, will be of considerable value, especially for innovation-financing, which the private sector typically perceives as high-risk. Hence, private 'proximity capital' can clearly be of great importance especially as lender–borrower interaction and open communication are seen to be increasingly important features in modern theories of finance. Thus, regional governance for innovation entails the facilitation of interaction between parties, including, where appropriate and available, the competences of member-state and EU resources. This can help build up capability, reputation, trust and reliability among regional partners.

However, regional *public* budgets are also important for mobilizing regional innovation potential. We may consider three kinds of budgetary competence for those situations where at least some kind of regional administration exists. First, regions may have competence to administer *decentralized spending*. This is where the region is the channel through which central government expenditure flows for certain items. Much Italian, Spanish and French regional expenditure is of this kind although there are exceptions, such as the Italian Special Statute regions and for some Spanish regions. A second category applies to cases where regions have *autonomous spending* competence. This occurs where regions determine how to spend a centrally allocated block-grant (as in Scotland and Wales in the UK) or where, as in federal systems, they are able to negotiate their expenditure priorities with their central state and, in Europe, the EU. The third category is where regions have *taxation authority* as well as autonomous spending competence since this allows them extra capacity to design special policies to support, for example, regional innovation. The Basque Country in Spain has this competence as will Scotland, to a far lesser extent, though Wales will not have this facility when its new National Assembly is established in 1999. Clearly, the strongest base for the promotion of regional innovation is found where regions have regionalized credit facilities and administrations with autonomous spending and/or taxation authority.

A further infrastructural issue concerns the competence regional authorities have for controlling or influencing investments in hard infrastructures such as transport and telecommunications and softer knowledge infrastructures such as universities, research institutes, science parks, and technology transfer centers. Most regions lack the budgetary capacity for the most strategic of investments, but many have the competencies either to design and construct many of them or to influence decisions ultimately made elsewhere concerning them. We broadly classify the enormous range of possibilities into types of infrastructure over which regions may have more or less managerial or influence capacity. Thus regions are likely to have no control regarding strategic investments such as a major international airport, some may have some control and influence over, for example, the provision of local or regional communications, some may be in a position to share control and management for, for example, regional science parks or research institutes, and they may have responsibility for the provision of, for example, technology-transfer centres. The greater the scope of competences with respect to the provision of hard and soft infrastructures for innovation, the greater the prospects, in principle, for the animation and facilitation of systemic regional innovation.

Superstructural issues

Three broad categories of conditions and criteria can be advanced in respect of superstructural issues. These refer, in general, to mentalities among regional actors or the 'culture' of the region and can be divided into the *institutional level*, the *organizational level for firms* and the *organizational level for governance*. Together, these help to define the degree of *embeddedness* of the region, its institutions and organizations. Embeddedness is here defined in terms of the extent to which a social community operates in terms of shared norms of cooperation, trustful interaction and 'untraded interdependencies' (Storper and Scott, 1995) as distinct from competitive, individualistic, 'arm's length exchange' and hierarchical norms. The contention here is that the former set of characteristics is more appropriate to systemic innovation through network or partnership relationships. This does not mean that innovativeness is not also associated with conditions of 'disembeddedness' since certain American cases such as Microsoft and Silicon Valley would appear to conform to that state. However, it should be noted that the work of Saxenian (1994) points strongly to the conclusion that a key reason for Silicon Valley's better long-term innovation performance than that of Route 128 Boston was that Silicon Valley was the region with the greater embeddedness.

Therefore, if we look, first, at the institutional level, the 'atmosphere' of a cooperative culture, associative disposition, learning orientation, and quest for consensus would be expected to be stronger in a region displaying characteristics of systemic innovation, whereas a competitive culture, individualism, a 'not invented here' mentality, and dissension would be typical of non-systemic, weakly interactive innovation at regional level. Moving to the organizational level of the firm, those with stronger systemic innovation potential will display trustful labor relations, shopfloor cooperation, and a worker welfare orientation with emphasis upon helping workers improve through a mentoring system, and an openness to externalizing transactions and knowledge exchange with other firms and organizations with respect to innovation. The weakly systemic firm characteristics would include antagonistic labor relations, workplace division, 'sweating' and a 'teach yourself' attitude to worker improvement. Internalization of business functions would be strongly pronounced and innovativeness might be limited to adaptation. Regarding the organization of governance, the embedded region will display inclusivity, monitoring, consultation, delegation, and networking propensities among its policy-makers while the disembedded region will have organizations that tend to be exclusive, reactive, authoritarian and hierarchical.

Regional competitiveness in Europe

Regional competitiveness is a concept with 'double articulation' in the sense that regions can be competitive on a 'low road' trajectory if they have low factor costs and 'good business climates', including low environmental, trade union and building regulation. Alternatively, they can be competitive on a 'high-road' trajectory of high skills, incomes, value-added, government enterprise support, innovativeness, infrastructural, and general economic efficiency. While we presume that competitiveness of the second kind is the type most favoured in the EU, we must recognize that there are regions whose characteristics approximate more to the first category within the EU. A key problem for them is that there are regions and countries elsewhere in the world, and nearby in Central and Eastern Europe, that are even better-equipped to offer 'low road' investment opportunities, so that option is not a particularly good one for the EU's less favored regions (LFRs) to pursue in any case. The Structural and Cohesion Funds and associated regional policies of the EU correctly recognize this fact.

In any case, it has already been noted in the introduction that 'the new regionalism' is a matter of recognizing and acting upon the theory that modern, competitive economic development rests increasingly upon the capability of regions to offer significant opportunities for firms both large and small, indigenous, and FDI, to gain from the external economies of embeddedness, mutual learning and opportunities for interactive innovation. This places responsibility upon regional governance organizations to be 'intelligent' with respect to their knowledge and information services, 'evaluative' with respect to understanding the fit between policies and objectives, and to have a 'monitoring' disposition towards the overall performance of the regional economy (Cooke and Morgan, 1998). The debate within Regional Policy DG16 of the EU, about whether the new regionalism should be matched by a 'new regional policy' which promotes innovativeness is an interesting reflection of processes of economic evolution (Landabaso, 1997).

But whether or not Structural Funds could usefully be used to promote and fund a greater emphasis upon the fashioning of 'soft infrastructures' in the way they have been used in the past to develop 'hard infrastructures' like roads and other basic facilities, depends to some extent, on the evidence that regions in the EU are meaningful economic entities with which policy of this more sophisticated type can sensibly interact and find some purchase.

One way of approaching this is to survey firms in EU regions to find out the nature and extent of firms' attachments to the region in which they find themselves. In the 'Regional Innovation Systems' study, firms in the EU regions demonstrated rather high functional and trading relationships within their region and member-state, as well as variable, but not insignificant interaction within the region for competitiveness-enhancing services and support associated with innovation.

The regions under investigation are diverse, being drawn from northern, eastern, southern and western parts of Europe, including two from outside the European Union, in Poland and Hungary, which are included in accession negotiations for the next round of EU enlargement. Nevertheless, on initial inspection, and primarily in terms of their economic structures and trajectories, they can be located in one of five broad categories.

Category 1 – High Performance Engineering: Southeast Brabant; Baden-Württemberg

Two regions, Baden-Württemberg (BW) in Germany and Brabant in the Netherlands are represented among the *high performance engineering regions*. High performance engineering takes place in large, indigenous firms with integrated supply-chains but some problems of competitiveness and possible future reconversion problems. In the case of Baden-Württemberg, this is because a major part of the regional economy is dominated by the automotive industry (Daimler-Chrysler, Porsche, Audi), and its suppliers, the electronics industry (SEL-Alcatel, IBM, Hewlett-Packard) plus supply-firms, and mechanical engineering (Trumpf, Traub, Heidelberg), especially machine tools and industrial machinery, for example, printing presses. In the case of Brabant, the regional economy is dominated by Philips and Daf along with numerous SMEs that act as suppliers to them or independently of them.

Category 2 – Reconversion with Upstream Innovation Emphasis: Styria, Tampere

This category refers to regional economies based on traditional industry in decline (coal, steel, shipbuilding, textiles) but which are engaged in *reconversion with upstream innovation* interactions, in which linkage to universities and research institutes to promote new industry clusters is pronounced. The two regions falling into this category are the Austrian *land* of Styria and the Finnish region centered upon the city of Tampere in southern Finland. Both are reconversion regions in receipt of European Union Structural Funds. There are similarities between these regions in that both are strong in the forest products or pulp and paper processing industries. But, importantly, both have the significant presence of firms that manufacture processing equipment for the pulp and paper industry and are, indeed, two of the world-leading locations for the production of this equipment. Both are also committed to university–industry interaction to stimulate the growth of new firms and sectors.

Category 3 – Reconversion with Downstream Innovation Emphasis: Basque Country; Wallonia; Wales

This category concerns regions displaying reconversion with downstream innovation trajectories whereby innovation impulses in developing clusters and/or sectors derive more from firms' roles in the supply-chain to larger customers. Three regions occupy this category, the Basque Country, Wales and Wallonia. Each suffered significantly from the decline of older heavy industry, notably shipbuilding, steel manufacture and coal-mining. Thus each region is a beneficiary of EU Objective 2 status and with this assistance, efforts have been made to restructure the regional economies toward the development of new sectors or clusters.

Category 4 – Regions with Industrial Districts: Friuli; Centro.

The fourth category concerns regional economies, peripherally located in the EU, with important clusters of SMEs in industrial districts contributing to key economic activities, typically in traditional industries such as furniture, leather and metal products. The two regions in question are located on the eastern and western edges of the EU, respectively Friuli-Venezia Giulia on the Italian–Slovenian border, and the Centro region, south of Porto in Portugal. Both are characterized by diffuse SME and micro-firm production systems focused upon traditional industrial sectors.

Category 5 – Transitional economies: Féjer, Lower Silesia

The fifth category refers to transitional economies, the two Central and Eastern European regions of Lower Silesia in Poland and the Féjer region, west of Budapest, in Hungary. Both exist in former state-controlled governmental systems in which regions had no meaningful place and innovation was state-directed, often to military ends. Now, there is more awareness of the importance of regional proactivity in a globalizing economy where FDI is seeking new markets and production locations. These regions, like their countries, are in transition, and are important centres of electrical and mechanical engineering, Féjer region having major western FDI from the likes of General Electric, IBM, Audi and Opel.

The upstream–downstream distinction for reconversion regions signifies whether major interaction by firms looking to become more competitive with innovation support occurs upstream towards knowledge-centers, or downstream towards the market via customer demand. The industrial sectors (sometimes also 'clusters' in the Porter sense (1990)) focused upon were mostly in manufacturing industry or services closely linked to manufacturing. This was because most innovation support tends to be directed towards manufacturing because it continues to play a key role in GDP exports even though it mostly plays a lesser role in employment.

If we look first at the decision autonomy of multiplant firms within their region for R&D (a very high order business function), the origin and destination of inputs and outputs, the location of their main competitors, and the location of firms with whom a cooperative partnership on any aspect of business functions is to be found, we find that regional focus is rather pronounced for key business activities (Table 9.1). It is generally the case, where data permit, to observe that the stronger regional economies such as Brabant and Baden-Württemberg more generally have a regional business focus at a reasonably high level across more business practices (or perceptions, in the case of recognition of where main competitors are) than the reconversion or industrial district regions.

Table 9.1   Percentage of Multiplant Firms in Key Manufacturing Sectors Reporting Regional Business Focus

Region	R&D conducted (%)	Inputs purchased	Outputs sold	Main competitors	Main cooperators
High performance engineering					
SE Brabant	28	38	35	42	NA
Baden-Württemberg	83	44	28	65	49
Reconversion (upstream)					
Styria	50	21	19	27	47
Tampere	33	28	26	30	44
Reconversion (downstream)					
Basque C.	58	52	28	40	59
Wallonia	34	26	25	21	30
Wales	47	22	20	21	32
*Industrial districts**					
Friuli	27	22	12	45	40
Centro	55	18	17	18	39

Source: REGIS Survey Data
* All Firms

Anomalies occur, of course, such as Brabant's lower R&D than most other regions and the Basque Country's higher than normal regional economic introversion, not unconnected with its industrial and political history. Moreover, relatively high scores for conducting R&D by regional firms need to be treated with caution since they register the proportion of firms that conduct R&D in the region, rather than representing the proportion of R&D being conducted by all firms in the selected sectors or clusters. Finally, the interpretation of R&D developed from follow-up interviews with representative firms suggests strongly that much of it is, of course, incremental development work rather than applied research, particularly in the industrial districts. Many qualifications! However, there are less with respect to the other indicators and what we see are far from negligible regional interactions for firms in key manufacturing industries, even in economies that have suffered from problems in their traditional industries. This should not surprise us since the majority of firms interviewed comprise small and medium enterprises for whom regional and national markets still reign supreme. Nevertheless, many of these are in supply-chain relationships to larger firms and, in turn, these are likely to be engaged in larger-scale competitive struggles. So global forces can react back significantly upon smaller, more regionally focused producers.

It is in this respect that the competitiveness judgements and perceptions of surveyed firms across a wide variety of regional and member-state settings are so revealing, especially considering also the variety of manufacturing sectors to which they belong. In Table 9.2 a summary of firms' views of their competitive advantage, how they plan to sustain it, the competitive advantage of their main competitors, and their response to that competitive challenge reveals the nature and apparent universality of the current 'drivers' of competition and the force behind the imperative to innovate. We see, very clearly, a broad statement of the nature of competition within the EU. Most firms rank 'quality' as the main selling point or competitive advantage with which they face the market. Even in Baden-Württemberg and the Basque Country where 'innovativeness' and 'delivery time' score highest, 'quality' ranks a close second. Most firms then rank 'skills' (and knowledge levels of their labor force) as the most important strategy to develop in order to sustain their present competitive advantage.

Where improving internal R&D or firm organization comes first, most regional firms rank skills enhancement second or third in importance, ahead of patenting, marketing, interfirm cooperation or support of other institutions.

Table 9.2 Main* Competitiveness Practices and Challenges of Regional Firms (1996)
*Likert Scale 4–5 Responses

Region	Main competitive advantage	Sustain competitive advantage	Competitive challenge	Competitive response
High performance Engineering				
SE Brabant	Quality (89%)	Skills (94%)	Quality (72%)	Cut cost (68%)
Baden-Württemberg	Innovation (86%)	R&D (68%)	Price (77%)	Cut cost (74%)
Reconversion (upstream)				
Styria	Quality (82%)	Skills (70%)	Price (76%)	Cut cost (62%)
Tampere	Quality (78%)	Skills (83%)	Quality (59%)	Organization (57%)
Reconversion (downstream)				
Basque C.	Delivery Time (50%)	R&D (39%)	Price (63%)	Cut cost (58%)
Wallonia	Quality (73%)	Skills (87%)	Price (70%)	Cut cost (69%)
Wales	Quality (84%)	Skills (70%)	Price (70%)	Organization (61%)
Industrial districts				
Friuli	Quality (86%)	Skills (67%)	Price (75%)	Organization (53%)
Centro	Quality (89%)	Organization (65%)	Quality (74%)	Cut cost (62%)

Hence human capital is ranked above social capital in the struggle by firms to retain competitive advantage.

However, when we examine what firms perceive to be the main competitive challenge they presently face, we see that price-competition generally comes first. Where it is transcended by quality-competition, price-competition ranks a close second in all three regions. There is a noticeable unanimity in the proportion of surveyed firms placing price-competition first (mostly 70–77 per cent). But there is also unanimity for firms assailed by price as a competitive weapon regarding what ranks next most importantly. In the majority of cases, quality is the next most highly ranked variable.

Finally, how do firms propose to respond to the price and quality threats of their competitors? Mainly it is to 'cut costs'. Even for those regions where firms stress the primacy of initiating 'organizational change' in the firm, they usually place cost-cutting as the second priority. The other expressed intentions scoring relatively highly are to intensify internal R&D and 'speed-up product development' in more or less equal measure.

Hence, the 'drivers' of innovation are the twin imperatives of raising the quality of products or processes produced and lowering their costs. To maintain quality, firms recognize the centrality of enhancing human capital, yet in the face of cost-cutting by competition, firms in a considerable variety of EU regions propose – inevitably – cost-cutting, but also some improvement in business organization and, to some extent, further investment in R&D. Two additional things should be noted at this point. First, most of the firms in the survey are SMEs, many in supply-chain relationships with customers who are demanding quality and cost improvement as condition of supply contracts. Second, firms internalize the quest for innovativeness to a very high degree, ranking cooperation with others as a solution relatively low in their hierarchy of imperatives.

Results for the two Central European regions involved in the study are quite similar to those for regions in the EU. Thus, in Féjer region in Hungary, located west of Budapest, 'quality' (70 per cent) just outscores 'time of delivery' (68 per cent) as the highest ranked competitive advantage. In both cases 'skills' is the key mechanism for sustaining competitive advantage, at 68 per cent in Féjer and 86 per cent in Lower Silesia, followed by 'innovativeness' and 'marketing' respectively. The main competitive challenge in Féjer is 'price competition' (78 per cent) followed by 'requirement of increasing product quality' (71 per cent), while in Lower Silesia it is 'price competition' (72 per cent) first, followed by 'organizational restructuring' (54 per cent, Féjer; 60 per cent Lower Silesia).

Innovation practices of European firms

Firms were asked a battery of questions about innovation. First, the question of organizational innovation was asked to ascertain the extent to which

Table 9.3 The Six Most Common Organizational Innovations of Regional Firms in Europe

	Organizational Innovation (% of Firms)						
Region	ISO 9000	Team-work	Total quality manage-ment	IT systems	'Just in time'	Flat hier-archy	Total
High performance engineering							
SE Brabant	24	57	46	24	40	51	40
Baden-Württemberg	68	49	51	25	24	73	48
Reconversion (upstream)							
Styria	60	59	36	51	37	71	52
Tampere	52	54	39	67	28	45	47
Reconversion (downstream)							
Basque C.	69	54	59	60	43	11	49
Wallonia	54	53	59	25	68	11	45
Wales	70	28	51	60	45	33	48
Industrial districts							
Friuli	42	62	59	63	53	49	55
Centro	54	47	65	30	33	28	43
Transitional regions							
Féjer	44	59	25	43	21	32	37
Lower Silesia	18	46	40	42	58	28	39
Total	50	52	46	45	41	39	46

modern management techniques such as Total Quality Management (TQM), Group or Team Working and System or Modular Supply (as distinct from discrete parts or components supply) were practiced. Thereafter, with a view to seeking information that would indicate the nature and extent of *interaction* regarding innovation, questions were put concerning sources of innovation knowledge and information and main partners in innovation. Questions were also asked about R&D expenditure as a percentage of turnover and staff in the knowledge that, while R&D statistics do not necessarily measure innovation, they do indicate an important commitment to innovation on the input side. Firms were asked to state their product and process innovations, new to the market and new to the firm, and also to identify the main constraints experienced in pursuing innovation aims.

Regarding organizational innovation, first, it is interesting to note that *team-work* is the most widely introduced organizational innovation. Though only just over half the sample had introduced it, it has been more or less equally spread across the regions despite their varying economic situations.

Secondly, *quality*-oriented organizational innovations such as ISO9000 and TQM have been introduced to a nearly equivalent degree to team-work, reinforcing the widespread claim that *quality* is seen as the main competitive advantage. Moreover, TQM is most widely practised in the regions which probably have more market-driven supplier relationships with major customer firms, for example, the 'downstream reconversion' and 'industrial district' regions. And thirdly, there tends to be a rather sharp decline in the adoption of 'flattened hierarchies', a fairly sophisticated management tool, outside the 'high-performance engineering' and 'upstream reconversion' regions (with the exception of Friuli) suggesting, perhaps, that cost-cutting by reducing management hierarchies has proceeded further in the higher labor-cost regions. On average the organizational innovation with lowest uptake was 'system supply', which was only practiced by 21 per cent of firms – the 'industrial districts' regions were exceptions with 45 per cent (also Lower Silesia). Other generally weakly practiced organizational innovations were 'interdisciplinary design' (simultaneous or concurrent engineering) at 22 per cent and 'interorganizational networking', also 22 per cent though, again, Friuli and Centro were higher than many on these 'associative' practices.

Moving to *sources* of the awareness among firms of innovation information, the top four sources are as presented in Table 9.4.

Table 9.4   The Four Most Important Sources of Innovation Information

Region	Journals	Conferences, fairs, exhibitions	Customers	Suppliers	Total
High performance engineering					
SE Brabant	28	23	48	24	31
Baden-Württemberg	52	71	71	26	55
Reconversion (upstream)					
Styria	55	57	55	50	54
Tampere	19	25	29	17	23
Reconversion (downstream)					
Basque C.	20	40	44	25	32
Wallonia	22	27	16	20	21
Wales	51	42	46	39	45
Industrial districts					
Friuli	23	29	23	23	25
Centro	38	66	31	35	43
Transitional regions					
Féjer	30	35	51	34	38
Lower Silesia	88	85	61	55	72
Total	39	46	43	32	40

With respect to the testing of systemic interaction in the quest for sources of information concerning innovation, it is evident that the German and Austrian regions are in some ways open to public domain sources such as journals, conferences and so on, but also rather closed in their scope of sources with respect to firms. Both are rather heavily reliant on their customers for innovation information, though rather less on their suppliers. Styria is the region in the EU with most dependence on supplier-firms for innovation information (except Lower Silesia, which is astonishingly high, for unexplained reasons, on all dimensions). There is no strong regional pattern; regions in the same category often perform very differently in respect of acquisition of innovation information. The strong conclusion here, echoed in comparable surveys, is that firms learn most from other firms with whom they interact in market transactions involving innovation and its associated information exchange. If we add in other possible sources such as, consultants, industry associations, technology transfer centers, universities and other educational bodies, most regions score mainly in single figures, that is, very low percentages of regional firms learn from these sources. The slight exceptions are industry associations in Brabant, BW, Wales and Centro, and universities in Styria and Wales where learning at twice the average rate is registered. Mainly, though, the 'soft infrastructure' of innovation support is weakly used for interactive innovation compared to journals, conferences/fairs/exhibitions, and firms in the supply chain.

Further investigation of more systemic linkage in the innovation process revealed very convergent results. In every region the main innovation partner is usually the customer firm, with supplier firms second. This ranking is reversed only for Wallonia, Friuli and Centro. At third place are universities and consultants equally. There is no evident pattern, in terms of regional economic character, in these third-choice partners. What is slightly surprising is the appearance of consultants, given their low average visibility as sources of innovation information. This may be interpreted to mean that firms use consultants for expertise in problem-solving rather than as sources of innovation ideas. The geographical location of customer and supplier partners in innovation is predominantly national followed by European Union-based and regionally based. The use of universities tends to be regional, then national, while the use of consultants is predominantly regional. Thus, a picture is reinforced of regional firms operating primarily on a regional or national scale with respect to interfirm relations in rather the same way that they do with respect to inputs and outputs (Table 9.1). The lesser use of innovation partners regionally contrasts with the higher levels of interaction with regional partners for non-innovation activities, as shown also in Table 9.1.

Although R&D expenditure is a poor measure of innovativeness, it is, as already noted, an indicator of commitment to an annual input of investment relevant to future innovation. In the regional firms surveyed and for the mainly manufacturing sectors focused upon, the overall average R&D investment for 1995 was 3.9 per cent of firm turnover, while the range was from 10 per cent in the Basque Country to 0.1 per cent in South East Brabant. In the first Community Innovation Survey (CIS) of 'innovation intensities' measured by current and capital expenditures on innovation activity as a proportion of turnover of firms in Norway and ten EU member states in 1992, the average was 7.2 per cent, ranging from 10 per cent in firms employing less than 50 persons to 4.5 per cent in firms employing more than 499. Three things may account for this: first, regional sectors were more representative of low and medium-technology industries (most of the nine EU regions are recipients of Structural Funds); second, response rates were better from medium or larger firms (certainly true in the 'industrial districts' regions) who spend a lower share of turnover on R&D; finally, we also have the CEE regions in the survey, and they lower the average considerably.

Hence, we are discussing what may be a more representative picture of European innovation capacity than that presented in the CIS data. Anyway, the regions in question tend to be those with problems of reconversion and development and are therefore the kind of economies specifically targeted by EU Structural Funds and Regional Policies. Given this, it is important to establish the extent to which firms in such regions have innovation potential and what their practices are in seeking to realize that potential. On some of the chosen indicators, we are again able to compare with CIS survey data. Firms were asked if they had introduced products and processes new to the firm and new to the market between 1993 and 1996. The results are presented in Table 9.5. They show three key things. First, product innovation inside regional firms is rather high in most EU regions and the two CEE regions, but product innovations new to the market are much lower, only about two-thirds reach the market. The figure of 44 per cent of firms introducing product innovations new to the market 1993–96 is similar to that found in the CIS Study 1990–92 where 48 per cent of firms in seven EU countries reported new products in sales. Thus an average statistic of around 46 per cent of firms being product innovators seems a sound one. Second, process innovations are less common, though an average of 50 per cent of regional firms reported them. However, less than half of these firms (23 per cent on average) take their process innovations to the market. The 'industrial district' regions tend to be the most innovative regarding processes new to the market while Baden-Württemberg is the most innovative with respect to placing new products on the market. Third, clearly, European firms engage in a considerable amount of innovative activity which does not result in a return on investment made, at least in direct terms. However, it appears, on the basis of face-to-face interviews with a representative sample of regional firms that much unsold process innovation occurs in order to achieve product innovation. Indeed, the two often go hand-in-hand.

Table 9.5 Product and Process Innovations 1993–96 and CIS Comparison

Region	Innovation (% of Regional Firms)			
	Product	New to market	Process	New to market
High performance engineering				
SE Brabant	36	17	28	8
Baden-Württemberg	79	63	39	13
Reconversion (upstream)				
Styria	67	48	44	21
Tampere	76	44	51	43
Reconversion (downstream)				
Basque C.	66	26	52	12
Wallonia	74	43	41	17
Wales	64	45	52	20
Industrial districts				
Friuli	80	51	76	26
Centro	83	52	74	26
Transitional regions				
Féjer	59	40	51	19
Lower Silesia	58	55	47	43
Totals	67	44	50	23
EU–CIS 1990–92 (New Products in Sales)	NA	48	NA	NA

Firms were further interrogated regarding the constraints or barriers they experienced in seeking to produce innovations. As can be seen from Table 9.6, which lists only the five most frequently cited constraints on average across the regions, 'funding' comes first, marginally ahead of 'management time' and the 'costs of research personnel', followed by lack of appropriate 'skills' and 'know-how'. Unimportant factors are 'market information', 'finding sources of know-how', and 'finding specialists' to assist in innovation activities. Thus firms are hampered mainly by internal rather than external factors. These devolve into funding and costs of personnel, management time, skills availability, and know-how issues. These results are not inconsistent with those of the CIS findings, though, there, the magnitude of the barrier constituted by inadequate finance is much higher.

But if 'funding' and 'costs of research personnel' are combined, the financial barriers to innovation are much closer (59 per cent compared to 55 per cent). 'Management time' and 'skills' together also come close to the CIS 'competence' barrier (42 per cent compared to 36 per cent on average). Variations between the regional categories show the 'reconversion' regions having the higher 'know-how' barriers and the 'transitional regions' the higher 'funding' constraints on innovation.

Table 9.6   Constraints on Innovation, 1996 and CIS Comparison

Region	Innovation Constraint (% of Firms Stating)				
	Funding	Know-how	Management time	Skills	Research personnel costs
High performance engineering					
SE Brabant	27	10	39	12	20
Baden-Württemberg	25	9	17	20	40
Reconversion (upstream)					
Styria	40	26	38	19	40
Tampere	21	18	30	13	19
Reconversion (downstream)					
Basque C.	29	15	21	15	18
Wallonia	40	11	26	25	13
Wales	18	18	35	31	17
Industrial districts					
Friuli	18	7	14	16	20
Centro	26	16	9	21	26
Transitional regions					
Féjer	48	9	10	13	39
Lower Silesia	78	9	16	20	24
Totals	34	13	23	19	25
		(Information)	(Competence)		
EU–CIS 1990–92	55	27	36		NA

Finally, firms were asked about cooperation in relation to innovation. The key hypothesis here is that cooperation between firms and among them and various elements of the innovation support infrastructure are signs of systemic product and process innovation, a factor that is reinforced to the extent that such cooperation is strongly regional, or (from a national innovation systems perspective) national. The results of responses to this question strongly confirmed those in Table 9.4 concerning cooperation partners for innovation information sources. Here are some key findings:

- Baden-Württemberg is entirely distinctive in the intensity of its firms' innovation cooperations with *customer* firms at the regional level (80 per cent of respondent firms having such links), at the national level (84 per cent), at the EU-level (65 per cent), and with the rest of the world (39 per cent). Brabant firms have their main innovation partnerships at regional level with *technology-transfer* agencies (40 per cent of firms have such links), at the national level with contract research organizations (13 per cent), at the EU level with universities (7 per cent), and they have no rest of the world innovation partnerships. Surveyed Brabant firms are all SMEs.

- Upstream reconversion regions have their strongest innovation partnerships with national *customers* (Styria, 72 per cent; Tampere 66 per cent) followed by regional customers (Styria, 54 per cent; Tampere 37 per cent), and thereafter by EU customers (Styria, 20 per cent; Tampere, 39 per cent) and customers from the rest of the world (Styria, 11 per cent, Tampere, 22 per cent).
- Downstream reconversion regions have their strongest interactions with national customers and suppliers (average 42 per cent of firms), followed by EU customers and suppliers (average 35 per cent), and then followed by global customers (average 19 per cent of firms). But their strongest innovation partnerships at regional level (mentioned by 44 per cent of firms on average) are with government agencies for grants and technology-transfer services
- The 'industrial district' regions have their main innovation cooperations with national (45 per cent) and EU (40 per cent) customers or suppliers, followed at a lower intensity by interactions with customers from the rest of the world (15 per cent). At regional level 23 per cent of firms have innovation links with customers or suppliers.
- In the 'transitional regions', customers or suppliers again dominate, 67 per cent of Lower Silesian firms engage in innovation partnerships with customer firms. This is made up of 32 per cent regional, 21 per cent national, 12 per cent EU, and 2 per cent in the rest of the world. Féjer region has a comparable pattern but, due to the strong presence of foreign multinationals, 24 per cent of innovation interactions are global customers, 50 per cent are EU, and 70 per cent of firms have both regional and national innovation partnerships.

Although the CIS data do not differentiate the spatial level of cooperation engaged in by product innovators, it is striking that for eight EU member-states plus Norway the average share of products obtained with technical cooperation with external partners by companies innovating in 1992 was 71 per cent. It is noted that in most EU countries incremental innovation involving technical cooperation with an external partner accounts for a large share of sales. Our results suggest that such partnerships are overwhelmingly with customer, and to a lesser extent, supplier firms and that the type of region mainly affects whether or not such partnerships are mainly national or regional, and in the latter case, whether they are public partnerships or not.

When firms do have interactions with the public innovation infrastructure, something that most in fact do engage in for innovation purposes, though not as a priority, the most commonly cited partner is the regional university system, closely followed by the national universities. Other regularly used services are those of regional, and to a much lesser extent, national technology centers or transfer agencies and venture capital sources, first

at the national level and then regional level. Contract research organizations at the national then regional level are also quite widely engaged in
support of firms' innovation activities.

When firms were further asked about the impact of using such services on
their capability, the most common response from firms in all regions is that
it 'speeds up the product development process' and 'enlarges the firm's technological base'. Less important impacts were those on skills, and collaboration with other firms or encouragement of wider collaboration with R&D
centres. Once again, firms use such services for internalized solutions to
problems arising in the course of conducting their own business activities. If
firms were not participating in publicly funded research projects or, more
generally, interacting with the innovation infrastructure, it is typically
because of 'no need to', 'internal solutions adequate', 'lack of information
about services', or 'bureaucratic application processes'. Cost was not usually
a major consideration nor was 'risk of losing know-how', though 'quality of
services on offer' was a relatively important reason for not using the services
of innovation organizations. These findings were broadly common for all
types of region, though the ability to access information of a technical
nature was a key reason for using services in the less accomplished regional
economies.

Policy issues for regions

The portrayal of competitiveness and the role of innovation in promoting it
in European regions is remarkably clear. European firms, in general, compete
on quality while they are competed against on cost. As unbiased observers,
we must temper this rather paranoid view by adding that since most of the
cost-competitors are European they must also be competing on quality too.
But moving beyond the therapeutic level to that of treatment, in what ways
can a role for policy be identified, assuming that policy has any justification
in any case? There is justification for policy support for regional firms in
Europe, not least because these firms both report and reveal the effects of
market-failure regarding innovation and competition. They report it when
they say 'funding', 'management time', 'skills', and 'personnel costs' are constraints on innovation. They reveal it when, in a collective way, they show
lack of competitiveness, innovativeness and job-generation capacity when
compared with the USA and Japan. But they do so, more specifically, when
they state that their customer or even supplier is the main source of the
information they draw on for innovation. Similarly the difference between
innovations and innovations new to the market is revealing. Also European
introversion and the belief that firms can solve their own problems (even
though innovation information comes from outside) suggests that they are
often incapable of doing other than taking instructions rather than setting
the agenda.

If European firms appear timid, then the organizations that apparently exist to help them innovate and become more competitive do very little to reduce their fears. Clearly, these organizations, into which substantial amounts of funding have been invested to employ significant numbers of public and private 'consultants', are not working. They are fragmented, over-lapping, confusing, bureaucratic, and, perhaps worst of all, perceived to be insufficiently skilled and expert to meet firms' needs. This is not surprising. Such organizations are new and inexperienced, their staff are often untrained in innovation support, they do not have much real responsibility or control of budgets, and they can only ever work with a few, favored firms. They, mainly, do not even have the resources or the incentives to find out from their market what its real needs or experiences of market-failure in fact are. The very recently introduced and very thinly spread EU support for Regional Innovation Strategies recognizes this but does not enable those regions fortunate enough to secure such projects to do very much to implement funded solutions to problems identified. Let us be clear here – without EU thinking and pilot project funding there would be almost no innovation-related enterprise support in less favored regions. The need, therefore, is not to destroy the relatively little good that comes from such efforts but to build upon them in a major way and make sure that funding is available from regional, national and EU levels to finance policies that meet actual firm needs.

This means radical change in regional innovation support funding. In a recent book, we argued that large firms should cease to be the main recipients of EU Framework Funding (Cooke and Morgan, 1998). This was because, in cases like Olivetti and Bull, both of whom have been major recipients, the money was wasted and for Siemens, it is a drop in the ocean compared to their vast internal R&D budgets. Further, we argued, backed by studies such as that of Malerba (1993), that large firms who are major recipients of national R&D funding from the public purse may not be innovators while those that receive nothing, at least in Italy, are both SMEs and modest innovators. Finally, supported by work such as that of Edquist (1997) we said that large firms who receive public R&D funding are often in declining sectors, so even if they innovate, the impact will be negligible. Regions cannot change this waste of resources themselves, but they are the main, strategic level of governance in close touch with SMEs. The most important actor who can change the present, massive imbalance is the EU itself. This is because, unlike most member-states, the EU recognizes regional innovation as an important issue and funds various programs in support of that view. The problem is that it mainly gives money for R&D to large firms who do not need it or cannot make use of it for serious innovation efforts.

So, starting at the EU level, future funding for science and technology should be allocated, through a competitive bidding process, to networks of firms, mainly small, supported mainly by universities, research institutes, or highly rated (by firms) technological centers. A large firm that can show that its involvement in the network is relevant or whose presence in the network is demanded by SMEs may be a network-member and recipient of an appropriate share of the project funding. Networks of firms must show that they have already experienced collaboration, rather than simply coming into existence for the purpose of bidding. In this way funding will go to those who most need it and who are likely to contain the most innovative potential. By encouraging firms into a more cooperative stance they will be made more efficient because of learning gains from each other, thus reducing the need for them to keep on wasting resources by constantly 're-inventing the wheel'. Maybe it will help them get more objective innovation information than they now rely on from their 'closed system' of main customer and suppliers. Universities are the most used support element in many regions and their involvement will help force them to become more relevant to the wider community in at least some of their research activities.

But what of the regional governmental level? We have seen how, in a remarkably straightforward way, substantial R&D resources can be diverted to those who can make most use of them. There are plenty of things that can be done and done better to help, but again the EU is potentially very important as an *animator* – more so than many member states. One of the most useful starting points is to conduct a Regional Innovation Strategy (RIS) project. Regions often wait to be given the idea and opportunity of funding to do such a thing, but why? These are not particularly expensive exercises, regions could easily fund them themselves, and some, like the Basque Country, have already done some of the strategic innovation analysis involved in RIS-type work, three times at least. But the problem is that the performance of RIS is only a beginning. It tells you what innovation assets the region and its firms possess, it enables the regional authorities to bring actors together to develop a consensus on future strategy, it even enables concrete projects to solve existing problems or meet present needs to be designed. But can they be implemented?

One way they can be implemented for less-favored regions (LFRs) is for RIS outputs to be 'mainstreamed' into Objective 1, 2 and 3 Single Programming Documents for future regional bids for Structural Funds. This should be the EU's 'new regional policy' approach. The EU is correct in shifting, or seeking to shift, its infrastructural spending in the 15 member states towards innovative 'soft' infrastructure such as enterprise and innovation support and away from road-building and other 'hard' infrastructure expenditure. This recognizes the gravity of the innovation gap in Europe and tries to help remedy it. This can be done without RISs, of course, but it is better for such program bids to be coherent and well-informed as well as supported by a regional consensus. The key thing is for the EU to make it clear to LFRs that they want RIS-type analysis and ensuing project-proposals to be contained in requests for Structural Funds from 1999 onwards.

Some of the kinds of things such funding could be spent on would be aimed at assisting firms to overcome the major constraints on innovation that they currently experience. All have some sort of difficulty in accessing investment capital to develop innovation, so a major element should be the establishment of regional venture capital funds and networks (for example, networks of 'business angels'). Firms continue to lack 'know-how', so clearly both the market and the enterprise support systems at regional level are failing. Much more market-focused research needs to be done on this in an ongoing, repetitive way and solutions to know-how problems found. Skills shortages are a major problem. For instance, Europe is short of about 2.5 million software programers. There are many other technical skills shortages. EU-funding linked to national and regional programs targeted on this are vital. Management has insufficient time to focus on innovation. Why not relieve management of the burden of trying to do too many things inadequately, and subsidize (for one or two years) 'Innovation Assistants' in SMEs, to arrange the better coordination and networking focus that innovation requires. Germany, the Netherlands and parts of the UK such as Wales have experience of the successful operation of such a scheme, sometimes called 'SME Graduates', in which case the problem of graduate unemployment is also moderated.

This is just a taste of thinking and action that can flow from taking a more systemic, regional view of the innovation question. From pursuing a new regional policy which places SMEs and knowledge-centers at the heart of the innovation process, the EU and the regions together can take major steps in meeting expressed needs.

Conclusion

There are three key points that emerge from this account of innovation and competitiveness among firms in diverse parts of Europe. The first of these is that, despite the hype about 'globalization', the majority of European firms spend most of their time and energy operating mainly on regional and national markets. Regions, in particular, are rather important (to between 27 per cent and 83 per cent of EU manufacturers) for R&D-related matters. They are also, to a rather lesser extent, important for sourcing inputs for production, but the national level is more important for sales of outputs. Finally, and crucially, in an age of interactive innovation, regions are significant as the sites of main cooperative activities between firms. So the regional level is most important for innovation and both competitive and cooperative input-related interaction. We may say that the region is the heart of interactive innovation.

Second, firms innovate because to compete they must produce higher quality at less cost. This is the universal story. Firms are forced to innovate whether or not they want to, in order to produce better things more cheaply. This is why such high figures are recorded for the proportions of innovative firms. But much of this effort is wasteful and firms are both too introverted and dependent on the customer or supplier for ideas. They need to be brought out into the open much more.

Finally, the organizations that exist to help firms to innovate are failing to do so. They are not used and not respected by firms because they do not meet their needs or help them to identify their needs. The whole regional innovation and enterprise support system is in need of serious overhaul with refocusing from the EU, regarding Framework and Structural Funding and more innovative thinking and action on innovation from the regions.

Acknowledgments

This paper summarizes the findings of research conducted in the 'Regional Innovation Systems: Designing for the Future' (REGIS) Project coordinated by the author. This research was funded by European Commission DG12 under the Fourth Framework – Targeted Socio-Economic Research Program. The author wishes to thank the European Commission on behalf of the eleven teams involved for funding the project. Thanks are expressed to all teams through their project leaders: P. Boekholt (The Netherlands), G. Bechtle (Germany); E. de Castro (Portugal); G. Etxebarria (Spain); A. Kuklinski (Poland); M. Quevit (Belgium); C. Mako (Hungary); M. Schenkel (Italy); G. Schienstock (Finland) and F. Tödtling (Austria). The usual disclaimer applies.

References

Cooke, P. and K. Morgan (1998) *The Associational Economy: Firms, Regions and Innovation*, Oxford: Oxford University Press.

Cooke, P., M. Uranga and G. Etxebarria (1998) 'Regional Systems of Innovation: an Evolutionary Perspective', *Environment and Planning* A, 30, pp. 1563–84.

Crevoisier, O. (1997) 'Financing Regional Endogenous Development: the Role of Proximity Capital in the Age of Globalization', *European Planning Studies*, 5, pp. 407–16.

Edquist, C. (ed.) (1997) *Systems of Innovation: Technologies, Institutions, and Organizations*, London: Pinter.

European Commission (1994) *Competition and Cohesion Trends in the Regions*, Brussels: CEC.

—— (1995) *Green Paper on Innovation*, Brussels: CEC.

—— (1996) *First Cohesion Report*, Brussels: CEC.

Johnson, B. (1992) 'Institutional Learning', in B. Lundvall (ed.) *National Systems of Innovation: towards a Theory of Innovation and Interactive Learning*, London: Pinter.

Krugman, P. (1995) *Development, Geography and Economic Theory*, Cambridge: MIT Press.

Landabaso, M. (1997) 'The Promotion of Innovation in Regional Policy', *Entrepreneurship and Regional Development*, 9, pp. 1–24.

MacIntosh, E. and A. Francis (1997) *A Review of Business Process Research across the Process Industries Sector*, Swindon, Engineering and Physical Sciences Research Council; Website Access of Paper – http://www.gla.ac.uk/Acad/FacSoc/reports/process research. htm.

Malerba, F. (1993) 'The National System of Innovation: Italy', in R. Nelson (ed.) *National Innovation Systems: a comparative analysis*, Oxford: Oxford University Press.

Malmberg, A. and P. Maskell (1997) 'Towards an Explanation of Regional Specialization and Industry Agglomeration', *European Planning Studies*, 5, pp. 25–42.

Marshall, A. (1919) *Industry and Trade*, London: Macmillan.

Porter, M. (1990) *The Competitive Advantage of Nations*, New York: The Free Press.

Putnam, R (1993) *Making Democracy Work*, Princeton: Princeton University Press.

Saxenian, A. (1994) *Regional Advantage: Culture and Competition in Silicon Valley and Route 128*, Cambridge, MA: Harvard University Press.

Storper, M. and A. Scott (1995) 'The Wealth of Regions: Market Forces and Policy Imperatives in Local and Global Context', *Futures*, 27, pp. 505–26.

10
Regions as Laboratories: the Rise of Regional Experimentalism in Europe

Kevin Morgan and Dylan Henderson

Introduction

With some notable exceptions the regional development debate in Europe has been dominated by exogenous models to such an extent that 'development' tends to be conceived as something that is introduced to, or visited upon, less favored regions (FRS) from external donors, be they states or firms. Laudable though it was in social welfare terms, this kind of regional policy did little or nothing to stimulate localized learning, innovation and indigenous development within FRS. Indeed, the latter were considered, and sometimes considered themselves, as passive receptacles for decisions taken elsewhere by others. In some of the most intractable regions – like Sicily and Campania for example – the problems of development are compounded by a corrupt and fatalistic political culture, where people feel themselves to be the powerless victims of circumstance, where local action seems pointless because, as the greatest novel of the Mezzogiorno put it, everything changes only for 'things to stay as they are' (Lampedusa, 1960). In these inauspicious circumstances the most important item on the developmental agenda is political reform and institutional renewal.[1]

We raise these issues at the outset because the developmental programs we focus on in this chapter invariably stand or fall on their ability to build social capital, that is a relational infrastructure for collective action which requires trust, voice, reciprocity and a disposition to collaborate for mutually beneficial ends. This focus on relational assets is part of the 'institutional turn' in regional development studies, a conceptual shift which has been triggered in part by growing dissatisfaction with dirigisme and neo-liberalism, the classical development repertoires which sought to privilege either state-led or market-driven processes regardless of time, space and milieu. The institutional perspective eschews the bloodless categories of 'state' and 'market' in favour of a more historically attuned theoretical approach in which the key issues are the quality of the institutional networks which mediate information exchange and knowledge-creation, the capacity for collective action, the potential for interactive learning and the efficacy of voice mechanisms (Sabel, 1994; Amin and Thrift, 1995; Storper, 1997; Morgan, 1997; Cooke and Morgan, 1998; Maskell *et al.*, 1998; Amin, 1999).

To avoid confusion it is worth saying that the institutional perspective does not jettison states and markets from its analytical framework in favor of networks of association; rather, it claims that networks have the potential to make both states and markets more effective: in the case of the state by creating a more dynamic policy environment with which it can engage, and in the case of the market by rendering it less of a Hobbesian war of all against all in which firms are inclined towards opportunistic behavior. For all their differences, the classical repertoires of dirigisme and neo-liberalism are paradoxically at one in devaluing the significance of that panoply of intermediary institutions between 'state' and 'market', such as interfirm networks, trade associations, chambers of commerce, civic associations, regional development agencies, labor unions and the like. Prosaic as they might seem, these self-organized intermediary institutions have the potential to play a significant role in fostering learning, innovation and development among their respective members and within their respective regions because, taken together, they constitute the institutional basis for collective action.

One of the key questions in regional development today – a question which resonates for theory, policy and practice – is whether these interactive learning networks evolve organically, through the repeated transactions of firms and their cognate associations, or whether they can be constructed through judicious public policy. It goes without saying, perhaps, that core regions of the world economy are well-endowed with robust interactive learning networks, since this is one of the reasons why they became core regions in the first place. What is much more open to question, however, is whether FRS are able to craft such networks to promote endogenous learning, innovation and development.

It is not that FRS are without networks, but that these tend to be of a predominantly vertical and asymmetrical character, which render local institutions highly dependent upon state or corporate hierarchies, in contrast to the more dynamic, horizontal networks which tend to form around agents of broadly equivalent status and power (Amin, 1999; Morgan and Nauwelaers, 1999). That vertical networks are less helpful than horizontal networks in solving dilemmas of collective action and promoting localized learning is because:

> A vertical network, no matter how dense and no matter how important to its participants, cannot sustain social trust and cooperation. Vertical flows of information are often less reliable than horizontal flows, in part because the subordinate husbands information as a hedge against exploitation. More important, sanctions that support norms of reciprocity against the threat of opportunism are less likely to be imposed upwards and less likely

to be acceded to, if imposed. . . . In the vertical patron– client relationship, characterized by dependence instead of mutuality, opportunism is more likely on the part of both patron (exploitation) and client (shirking) (Putnam, 1993, p. 174).

In contrast to the classical repertoires of dirigisme and neo-liberalism, which have little or nothing to say about trust, voice and reciprocity – relational assets which lubricate successful networks – the institutional perspective insists that these intangible resources merit as much attention as tangible resources (Cooke and Morgan, 1998).

In the following sections we examine these issues from a number of different vantage points: in the second section we examine some institutional dimensions of regional development in Europe; the third section charts the rise of a new paradigm for regional innovation policy in the European Union (EU); the fourth section offers a preliminary assessment of this policy in practice, using Wales as a case study; and the fifth section concludes by distilling some of the wider lessons of experimental regionalism in Europe.

Institutions, innovation and regional experimentalism

Nowhere has the regional role in innovation been more forcefully championed than in the EU, where the European Commission is seeking to address not just the symptoms of uneven development (like high unemployment rates and low GDP per capita), but also its causes (one of which is the FRS' weak innovation capacity). The rationale for this strategy is not merely an attempt to redress inter-regional disparities in the Union, a social welfare goal; it is also an attempt to redress the EU's 'innovation deficit' *vis-à-vis* the US and Japan, a developmental goal. The Commission believes that, relative to its two principal competitors, the EU has a poor record in converting its scientific and technological expertise into commercially successful products and services, a problem it attributes to shortcomings in transferring knowledge from laboratory to industry, from firm to firm and region to region, a problem of networking capacity (CEC, 1993). Although this problem varies across countries and regions on the one hand, and industries and firms on the other, it is especially acute in the FRS.

Unlike previous EU regional development programs, which were inordinately biased towards supply-side initiatives, like infrastructural schemes for example, the new generation of regional innovation policies represents a radical departure in two ways: first, they address themselves to the demand-side problems of local firms and, second, they aim to tackle the problems of institutional inertia in the FRS by sponsoring a new, consensus-based process of interactive learning within and between the public and private sectors (Morgan, 1997). Elsewhere we have argued that these new regional innovation policies signal a serious attempt, on the part of policy-makers and practitioners alike, to address the problems which have been identified in the institutional perspective on regional development (Morgan and Nauwelaers, 1999). In particular these new policies seek to engage with the most obdurate institutional problems (like voice-formation, cooperation and collective action) which have exercised a long line of developmental theorists from Albert Hirschman onwards.

The art of voice

It is a measure of Hirschman's colossal achievement that, forty years on, his work continues to resonate in development circles around the world. The definition of development which he proposed in 1958 remains as important today, for both analysis and policy, as it was when it first appeared. In contrast to the narrow and static definition offered by prevailing neoclassical economic theory, Hirschman argued that development depends not so much on finding optimal combinations for given resources and factors of production as on calling forth and enlisting for development purposes resources and abilities that are hidden, scattered, or badly utilized (Hirschman, 1958, p. 5).

The great merit of this conception, which treats resources as latent and conditionally available rather than outright absent or scarce, is that it directs attention to the essential dynamic and strategic aspects of the development process. In contrast to theories which stressed the scarcity of conventional factors – like capital, education and entrepreneurial skills for example – Hirschman argued that these 'scarce' factors could be reduced to one, more basic scarcity, namely 'the basic deficiency in organization'. In particular, he identified a shortage of 'the cooperative component of entrepreneurship' which, among other things, involved the art of agreement-reaching, conflict resolution and cooperation-enlisting activity, all of which were important because 'the fundamental problem of development consists in generating and energizing human action in a certain direction' (Hirschman, 1958, p. 25). To unlock institutional inertia, and call forth untapped resources, this analysis highlights the need to look for what Hirschman called 'pressures' and 'inducement mechanisms' that might elicit and mobilize these resources for development purposes.

Although Hirschman was principally concerned with underdeveloped countries, his analysis remains pertinent to the 'innovation deficit' of FRS in developed countries, particularly as regards the role of agreement-reaching and cooperation-enlisting mechanisms. In the context of FRS, where firms tend to see themselves in atomistic terms and where, consequently, there are low levels of social capital (for example, norms of trust and reciprocity), the most significant innovation will be to develop voice-based mechanisms through which firms and public agencies can learn to cooperate to explore joint solutions to common problems.

This is easier said than done, of course, especially in an environment where low-trust and exit-centered behavior are the prevailing norms, provoking a vicious circle in which opportunism begets low-trust and this in turn reinforces opportunism. The literature on trust suggests that the parties to high-trust relationships enjoy at least three important benefits: (1) they are able to economize on time and effort because it is extremely efficient to be able to rely on the word of one's partner; (2) they are better placed to cope with uncertainty because, while it does not eliminate risk, trust reduces risk and discloses possibilities for action which would have been unattractive otherwise; and (3) they have a greater capacity for learning because they are party to thicker and richer information flows (Cooke and Morgan, 1998).

If we accept that trust does indeed confer these benefits, this raises a question which perplexes theorists and practitioners alike, namely: how is it secured in the first place? Most of the literature on trust seems to fall into two, equally inadequate, categories. The existence of high-trust relationships is either attributed to 'cultural norms', which assumes precisely what needs to be explained, or it is reduced to 'calculative action', which exaggerates the parties' capacity for gauging the precise benefits of cooperation in advance. A more useful approach would recognize that trust is neither an outcome derived from calculation nor a norm traced to culture, but a disposition which is *learned* and reinforced through successful collaboration, such that trust is a by-product of success rather than a precondition for it (Powell, 1996).

Building trust therefore requires a constant dialogue between the parties so that interests and perceptions can be better aligned. Here it is worth revisiting Hirschman's concept of voice, which he defined as any attempt 'to change, rather than to escape from, an objectionable state of affairs' (Hirschman, 1970, p. 30). Although exit and voice tend to be seen as alternative reaction mechanisms to institutional problems, Hirschman was quite clear that they could also be complementary, in the sense that the voice option is strengthened by the threat of exit. However, voice would seem to be the richer of the two mechanisms because 'once you have exited, you have lost the opportunity to use voice, but not vice versa; in some situations, exit will therefore be a reaction of last resort after voice has failed' (Hirschman, 1970, p. 37). While exit seems to be the less costly, and more flexible option, it is not without costs: for example, it is difficult to see how firms can forge interactive learning relationships with their suppliers if they are forever switching from one to another for the sake of short-term cost savings. In other words, the costs of voice, like reduced flexibility, need to be balanced against the benefits, like the greater potential for interactive learning.

Hirschman's original formulation has been rightly criticized for assuming that voice is a ready alternative to exit, a criticism he later accepted along with the charge that there are other reactions besides exit and voice, like passivity, resignation and withdrawal (Hirschman, 1986). Far from being readily

available, voice-formation requires investment, of both time and resources, if it is to be a viable alternative to exit. Whereas exit involves a clean break, voice is essentially 'an art constantly evolving in new directions' (Hirschman, 1970, p. 43).

Contemporary theories of learning, innovation and development continue to grapple with the problems identified by Hirschman, particularly the problem of voice-formation. Recent examples would be Sabel's work on 'learning by monitoring', which highlights the role of *discussion* as the process through which 'parties come to reinterpret themselves and their relation to each other by elaborating a common understanding of the world' (Sabel, 1994). Then there is Piore's hermeneutic perspective, where the problems of orchestration and cooperation are likened to a series of intersecting *conversations* in which the key issues are who talks to whom and what they talk about (Piore, 1995). Finally, Storper has emphasized the significance of 'soft' factors, like *talk* and *confidence*, in building the relationships of the institutionalized learning economy (Storper, 1997). Storper tentatively suggests that, under certain conditions, talk and confidence are 'more likely to succeed when they are geographically localized' and that small, repeated and low-cost experiments can help to induce interactive learning between parties in an environment which has hitherto been characterized by distrust or antipathy (Storper, 1997).

These issues lie at the heart of the regional innovation experiments which are taking place in Europe today, and these carry implications far beyond the confines of the regions directly involved in this process of experimentation (Sabel, 1995; Cooke and Morgan, 1998).

Devolution and regional experimentalism

Even if we accept that discussion, talk and conversation are important for building relational assets, this still leaves unresolved the question as to who is to assume the responsibility for animating these conversations and for deciding who talks to whom and what they talk about? In principle, this role could be performed by regional development agencies, chambers of commerce or even a large, locally hegemonic firm, though none of these organizations commands the political legitimacy nor the moral authority of a democratically elected regional government (Morgan and Nauwelaers, 1999). Piore is in no doubt where the responsibility for this task lies:

> The role of public policy and political leadership is to orchestrate those conversations, initiating discussions among previously isolated groups, guiding them through disagreements and misunderstandings that might otherwise lead conversation to break off, introducing new topics for discussion and debate (Piore, 1995, p. 138).

Although this assumes that the public authorities in the region are comp tent to fulfil this role of *animateur*, a truly heroic assumption in some FR.

nevertheless helps us to understand that devolution to regional tiers of the state is important for developmental purposes principally because it creates the opportunity for orchestrating more meaningful discussions and conversations between public and private sectors within the region. The regional level is now deemed to be an important arena in which to design and deliver policies to foster learning, innovation and development because it is said to be 'the basic level at which there is a natural solidarity and where relations are easily forged' (European Commission, 1995). While this may be broadly correct, the main problem with such a view is that it naively assumes that spatial proximity is a sufficient condition for forging interactive learning networks, when the evidence suggests that these have to be actively constructed through conscious and painstaking efforts on the part of firms and public agencies (Cooke and Morgan, 1998).

Suitably empowered, however, the regional level would seem to possess two potential assets which make it a suitable level at which to engage in the kind of support envisaged by the Commission. First, it is able to act on *local knowledge*, part of which is tacit, concerning the calibre of firms, the formal and informal linkages between them, the quality of the labor force and the capacity of the institutions which deliver technical, commercial and training services. Second, the regional level is perhaps the most appropriate for building *social capital* because this is the level at which regular face-to-face interactions – one of the (necessary) conditions for trust-building – can be sustained over time. The literature on trust and cooperation suggests that these are more likely to occur where there is a strong possibility that the agents will meet again, in other words where the shadow of the future looms large over the present (Luhmann, 1979; Axelrod, 1984).

What needs to be emphasized, however, is that these are potential assets which need to be mobilized through conscious political action because, contrary to what the Commission seems to think, they do not exist in some primordial form at the regional level. These assets will need to be utilized if FRS are to develop more effective systems of enterprise support services because, at the moment, they fail to touch some 80 per cent of the SME population in the EU (European Commission, 1995). Low service take-up is related to the fact that most services are delivered in a crude supply-side fashion by an enterprise support industry that finds it easier to offer a standard menu of services which has been designed for, rather than with, local firms. Indeed, in some cases the priorities of regional service providers are radically at odds with the priorities of the firms in the region, a fatal mismatch which suggests that the former are more interested in working to their own agendas.

The problems of supply-side bias and standard service menus also reflect the fact that public agencies in FRS often lack the skills to engage in interactive service provision, one reason why they have a credibility problem *vis-à-vis* the corporate sector (Morgan, 1996). The challenge of interactive service provision, in which the aim is to develop services with, rather than for, local firms

to enhance the latter's absorptive capacity, cannot be met through traditional supply-side regional policy; that is to say, technology centers and the like are not likely to resolve the innovation deficit in FRS if local firms are unable or unwilling to utilize these services – the 'cathedrals in the desert' syndrome. Hence one of the most important tasks in FRS is to stimulate conversations between providers and users to ensure that enterprise support services are driven by the requirements of local firms rather than the bureaucratic needs of the delivery agencies. Regional experimentalism, which aims to alleviate these problems by stimulating new and more purposeful conversations, can play 'an important part in overcoming the crisis of regional institutions that currently accompanies and echoes the confusion in the firms it is meant to address' (Sabel, 1995). To be effective, however, the lessons of regional experimentalism need to be acknowledged and disseminated by national and supranational authorities in the EU, a point to which we return in the final section.

Experimentalism in European regional policy

One of the clearest expressions of regional experimentalism in policy terms can be found in Article 10 of the European Regional Development Fund (ERDF). Despite accounting for less than 1 per cent of a total ERDF budget of 70 billion ECU for the period 1994–99, Article 10 funds have been responsible for some of the most innovative policy initiatives to emerge in Europe in recent years.[2] Unlike conventional mechanisms for distributing EU regional policy funds – development programs negotiated with Member States – Article 10 provides an opportunity for the EU to help establish its own innovative regional pilot studies. That is, Article 10 opens up the possibility for the European Commission to draw upon lessons from policy experiments across European regions and work directly with regional actors to explore new support measures. In addition, and perhaps most importantly, Article 10 also allows for a greater degree of *risk-taking* than is typically possible through mainstream EU regional development programs. This lack of innovation has been particularly evident in the Monitoring Committees charged with overseeing development programs at the regional level. For the most part the expertise of Monitoring Committees reflects the old regional policy era, when traditional infrastructure was the order of the day, rather than the new regional policy era, which is more concerned to promote innovation, the environment, human resources and equal opportunities.

It is worth remembering, however, that while Article 10 provides the means for the EU (in collaboration with regional institutions) to experiment with new policy initiatives, it does not do so in an overly prescriptive manner. Instead, Article 10 relies on the principle of helping regions to help themselves through initiatives designed to mobilize local knowledge in a process of collective social learning. The philosophy of Article 10 can therefore be summarized in the following terms (Messina, 1997):

- Article 10 is designed to be an experimental laboratory;
- Its aim is to promote the innovative dimension of regional policy;
- It seeks to promote partnership between the private and public sectors;
- It enables the internationalization of regions and local authorities;
- It aims to facilitate the transfer of know-how in the technical, economic and scientific fields between the regions of the EU; and
- The positive results of Article 10 projects should be incorporated in conventional regional policies.

In addition to these factors, it should also be noted that Article 10 not only provides an opportunity for the European Commission to engage in policy-related learning; its operating mechanisms also provide a powerful impetus for regional authorities to think strategically about the needs of companies and about the appropriate role for public sector intervention. In this sense then, Article 10 aims to set in train a series of interactive intra- and inter-regional learning processes. These features, we argue below, have been particularly evident in the recently established Article 10 program: Regional Innovation Strategies (RIS).

The origins of the RIS program can be seen in the context of the growing realization that Europe's most prosperous regions have succeeded in appropriating the overwhelming share of EU science and technology resources, the so-called Framework Funds. The Commission's own estimates, for example, suggest that some 50 per cent of all research and technological development (RTD) funds have been concentrated in just a small number of 'islands of innovation' – Amsterdam, Rotterdam, Ile de France, the Ruhr, Frankfurt, Stuttgart, Munich, Lyon, Grenoble and Turin (European Commission, 1996). The explanation for this disparity is to be found in the Commission's use of the principle of 'scientific excellence' – traditionally lacking in FRS – to guide the distribution of funds. EU policy, in this respect then, has tended to exacerbate rather than redress the socioeconomic disparities found in the EU.

In response to these problems, the European Commission's Regional Policy Directorate (DG XVI) launched the STRIDE (Science and Technology for Regional Innovation in Europe) initiative in 1990. In broad terms this program attempted to address many of the main barriers responsible for poor LFR participation in EU research programs. More specifically, STRIDE represented an attempt to build innovation and institutional capacity in FRS by providing assistance for strengthening technology, innovation and research infrastructure, participation in national and international RTD programs and cooperation between industry and RTD centers. STRIDE, however, was faced with a number of formidable obstacles to achieving this goal. In particular, follow-up assessments have found that the most peripheral (so-called Objective 1) regions were unable to mobilize sufficient private sector input into projects. Moreover, most sought funds for physical infrastructure at the expense of 'soft' measures intended to foster links or promote training. For

Landabaso (1997), the source of these weaknesses can be traced back to two main areas. First, many FRS lacked understanding of innovation as an interactive process. This led regions to overinvest in university 'science' projects, assuming that this would automatically feed through into the industrial environment. Second, organizations in FRS lacked the skills and expertise necessary for some of the 'network building' projects supported by STRIDE.

Despite its limitations, STRIDE is now recognized as an important learning experience by DG XVI (Landabaso and Reid, 1999). The Commission learnt, for example, that supporting innovation within FRS was not simply a matter of providing funds for RTD projects. Rather, they recognized that policies should focus on a more broad-based conception of innovation in terms of managerial skills, quality standards and organizational capacity. They also identified a greater need for new forms of institutional cooperation to design and deliver innovation policies adapted to local circumstances.

These insights formed the basis of the European Commission's successor to STRIDE, launched in 1994 under the name Regional Technology Plan (RTP). The main objectives of the RTP exercise were twofold. In the first place, it was designed to encourage FRS to develop a regional innovation process, in which the regional stakeholders were enjoined to define a commonly agreed, bottom-up strategy attuned to the nuances of their regions. Second, it was hoped that the RTP would provide a framework in which the recipient regions and the Commission could jointly agree on a more optimal strategy for future investments in RTD initiatives at the regional level.

The RTP was not, however, viewed as a one-off study; rather it was seen as first and foremost a strategy-making process which could foster dialogue between regional actors. In doing this, the Commission anticipated that the exercise might help regional institutions better understand the needs of firms and thereby

> gain experience of interacting between the business community, the public sector and the RTD community by means of stable, informal channels of contact through discussion groups . . . [thus] establishing a strategic planning *culture* at the regional level' (European Commission, no date).

In this sense, RTP was viewed as a vehicle not only for outlining a 'framework for decision making', but also to set precedents and build trust amongst regional stakeholders.

The practical content of the RTP strategy-making process, as set out in the Commission's guidelines, is summarized in Table 10.1. It should be noted, however, that these points are indicative; indeed the Commission strongly advised regions to adapt them to their particular circumstances (European Commission, 1997).

Table 10.1   Key Themes of the RTP Strategy Development Process

---

Theme 1: Assess the strengths and weaknesses of regional firms' RTD capacity and specify their needs, both expressed and latent, for innovation-related services.

Theme 2: Identify the key technological and industrial trends affecting the regional economy so as to determine threats and opportunities.

Theme 3: Analyze and assess the region's RTD supply-side infrastructure to determine the degree of 'fit' with the expressed needs of regional firms.

Theme 4: Define the orientations of the main institutional actors with the aim of creating a shared diagnosis of the problems and a widespread discussion of what priorities ought to be addressed in the RTP exercise.

Theme 5: Have the Steering Group, which should be drawn from as wide a constituency as possible, define and disseminate the strategic priorities of the RTP program and to explain how these are to be met in practice.

Theme 6: Establish a system for continuous monitoring and evaluation so that the main actors have regular feedback on problems as and when they arise.

Source: European Commission, 1994

The RTP and RIS exercises: collective learning in practice?

The RTP program was formally launched in June 1994. Eight regions were selected to pilot the exercise: Limburg (Netherlands), Lorraine (France), Saxony-Anhalt (Germany), Wales (UK), Castilla y León (Spain), Central Macedonia (Greece), Norte (Portugal) and Abruzzo (Italy). In 1996, the exercise was further expanded to include an additional 19 regions through an open call for proposals. At this point the exercise was renamed RIS in an attempt to encourage regions to pay more attention to the non-technological aspects of innovation. The budget for the RIS exercise now amounts to a maximum of 500,000 ECUs per region, of which 50 per cent is funded by the Commission. The main focus of the discussion below is to explore the extent to which the RTP exercise has been able to stimulate these collective learning processes in practice. In doing this, we draw on our experiences of the Wales RTP pilot, an exercise in which we participated as members of the Steering Group. We also provide some early indications of the relative progress of the Welsh exercise in the light of recent studies from other RIS regions (Tsipouri, 1998; Morgan and Nauwelaers, 1999).

For many years Wales has struggled to cope with the decline of its once dominant coal and steel industries. This period of industrialization was associated with high levels of social and environmental degradation in the region's industrial heartlands, and crucially, with an underdeveloped indigenous business class (Williams, 1980). In an attempt to redress these problems, regional policy has been in operation, with varying degrees of success since the early 1930s. A conspicuous feature of these efforts has been the attraction of a large number of inward investors. This strategy, in part, has helped to contribute towards the development of a more diversified economy based on manufacturing and services. Yet despite the recent success in encouraging investment from companies such as Bosch, Sony and LG, and the claims made by several commentators that the region is experiencing somewhat of an economic renaissance, Wales continues to suffer from a number of major structural problems. These include low economic activity rates, below average GDP per head, and a low position in the UK regional wages league table. In the light of these persistent weaknesses the Welsh Development Agency (WDA) and other support institutions have in recent years begun to experiment (alongside their long-standing inward investment activities) with a range of networking initiatives designed to increase the capacity of indigenous companies to innovate. A prime example of these new departures in regional policy is the EU-sponsored Wales Regional Technology Plan.

In many respects, Wales represents an important case study of the RTP repertoire, not least because it was the first region to petition the European Commission for a regional innovation program. The region was also actively involved in the early discussions which helped shape the practical content of the program (Henderson and Thomas, 1999). More recently it has been held up as a model of best practice by the European Commission (*see*, for example, European Commission, 1998; Arnold *et al.*, 1998). In short, the early stages of this process centered on three main research activities: *desk research*, bringing together various reports and papers on the Welsh economy, its innovative capacity and so on; 350 *Technology Audits* and a *survey of innovation and technology support infrastructure*. Together these elements provided the means by which the Steering Group was able to begin to develop an understanding of the main innovation issues facing firms in Wales. An important part of the process, however, was the 'testing' of these findings with regional firms and support organizations, and exploring, interactively, appropriate solutions.

Feedback from the early research activities was sought in two main ways. First, over 30 *Panel discussions* were held between February and November 1995 with representatives from industry, local government, higher and further education, schools, enterprise agencies, development bodies, trade unions and government. Panels were organized in consultation with regional support providers and brought together existing networks (client groups, for example). The objective here was to encourage *talk* around the key issues and trends facing Wales, and to identify potential responses. In addition to discussions, a special 'one-off' international experts meeting was held to review the RTP process, analysis and conclusions. This was viewed as an important mechanism to guard against parochialism, by seeking an outside perspective on the RTP. Many of the experts present at this meeting were, themselves, also involved in RTP exercises elsewhere in Europe. For this reason the panel helped to further support informal networking amongst RTP regions, as well as allowing participants to benchmark progress and exchange experiences. The value of this exercise has been highlighted by the fact that it is now part of the Commission's guidelines to RIS regions (European Commission, 1997). The second aspect of the consultation exercise in Wales was the production, distribution and presentation of a Consultative Report (WDA, 1996a). This outlined, in some detail, the main innovation issues, possible priorities and projects identified through the research and panel meetings. It was launched in January 1996 and circulated widely to firms, organizations and key individuals across the region. In response, well over a 100 organizations provided feedback.

The culmination of the relatively exhaustive consultation process – involving over 1000 organizations – was the launch of an Action Plan in June 1996 by the Secretary of State for Wales (WDA, 1996b). This set out details of six priority areas and some 66 'committed' projects where support had been obtained from regional organizations. Of these, the Plan designated a number of 'Flagship Projects', each associated with particular priority areas[3] (*see* Table 10.2). It was

Table 10.2 Priorities and Flagship Projects for the Wales RTP

- A culture of innovation is vital for personal and economic success

  The Welsh Innovation Challenge: a project to integrate existing innovation award competitions, providing a national profile with improved promotion, publicity and assistance with commercialization.

- Wales must profit from global innovation and technology

  The Welsh Optoelectronics Forum: a project to help expand the international networking activities of the region's optoelectronics group.

- Companies learn best from each other, therefore supply chains and networks are crucial

  Innovative Teaching Company Schemes: an extension of the successful Teaching Company Scheme (technology transfer from the university sector to industry via a two-year graduate placement) to enable companies in particular supply chains address a common innovation need.

- Finance for innovation must be readily available in Wales

  Technology Implementation Funding Program: a project to identify technology and innovation needs in SMEs and provide part-funding for the acquisition of new technology and consultancy for technical problem-solving.

- High quality business and innovation support is essential for Welsh companies

  Support Centers for Information Technology and Multimedia: establishment of centers to provide demonstrations of and access to information technology.

- Education and training for innovation and technology are vital for the Welsh economy

  Bargaining for Skills: a project to assist employers and trade unions in working towards training goals.

Source: WDA (1996a)

envisaged that these would be implemented in the period immediately follow-ing the launch of the Action Plan.

While the publication of the Action Plan marked the formal completion of the EC-funded exercise, many elements of the RTP process in Wales have continued to operate in the subsequent period. In particular, the Steering Group has remained in place to ensure that the momentum built up during the RTP was not lost. Perhaps the most significant activity undertaken by the Steering Group in recent years has been to 'revisit' the objectives and priori-ties set out in the original Action Plan. This monitoring and evaluation exer-cise, in part, emerged from the growing feeling that the original target of establishing 'a consensus on a strategy to improve the innovation and tech-nology performance of the Welsh economy' (WDA, 1998) had largely been met. It was also a product of the 1997 Devolution Referendum which result-ed in the establishment of the Welsh Assembly, and the decision to merge the WDA, Land Authority for Wales and the Development Board for Rural Wales in a new 'economic powerhouse' for Wales. These factors, the Group anticipated, could lead to the RTP being 'sidelined' to more immediate con-cerns within Wales's new institutional structures.

In response to these concerns the Steering Group established a new review process to provide the basis of a revised 'RTP 1998' which could determine progress and contribute towards this strategy debate. In a similar way to the consultation activities which supported the first Action Plan, RTP 1998 sought to generate an iterative and interactive process of discussion and knowledge exchange amongst a broad range of regional organizations. This incorporated a series of some thirty meetings with over 600 individuals, and largely confirmed the validity of the main priorities set out in Table 10.2 above (WDA, 1998). The review, however, also highlighted the need to fur-ther communicate the RTP objectives and priorities to the innovation and technology support community.

How, then, can we assess the Wales RTP exercise? With formidable objec-tives such as the generation of an innovative culture in Wales, it seems clear that many of the outcomes of the RTP Wales exercise are unlikely to be short-term in nature. A consistent theme from our research (Morgan, 1997; Henderson and Thomas, 1999) however, is that whatever the long-term impact of the RTP and its associated projects, participants feel that the strat-egy-making process itself has already been a valuable exercise. These benefits primarily fall into three main areas: acquiring a better understanding of the innovation process, best-practice support structures and the needs of firms; new interactions and relationships between regional support organizations; and more inclusive regional policy-making routines in the field of innovation.

Beginning, then, with the claim that the Wales RTP has been able to build new insights and awareness – this process was most evident in the use of interactive activities such as Steering Group meetings, panel discussions and so on, allowing many firms and support institutions to be exposed to new ideas about the needs of firms and the efficacy of different policy mechanisms. Engagement in the process also helped institutions to acquire important empirical confirmation for previous ideas. Participation in the RTP, however, did not simply produce 'cognitive' insights. Many organizations had already begun to *act* on these new understandings, translating them into new forms of organizational activities and behavior. This was evident in the way that RTP had encouraged the Welsh Office, WDA and DBRW to frame new strategy documents. The WDA, for example, have now taken the decision to use the six priorities of the RTP as a guide for its business development programs for the period 1998 to 2001. This is likely to ensure that future initiatives are designed with RTP priorities in mind.

Yet while it seems clear that the RTP has helped to raise awareness and understanding of the innovation process amongst innovation and technology support institutions in Wales, a key question yet to be resolved is whether the ideas contained within the RTP have actually been diffused sufficiently beyond the individuals and organizations represented at the Steering Group level. To answer this question fully would require more research; but, if nothing else, the RTP does appear to have forced many key regional actors to (re)consider their operational priorities in the light of the issues raised during the process. Whether this learning process ultimately brings about a business support system more in tune with the 'real' needs of firms, though, will be a product of the ability of Steering Group members and other regional actors to bring more depth to the process by communicating it further.

Moving on to the second main aspect of the RTP 'learning process' identified above: the creation of new interactions and relationships between regional support organizations. Here, RTP Wales made specific efforts to ensure a regional consensus by encouraging interactive dialogue between a wide cross-section of actors. Particularly important was the role of the Steering Group, which provided a 'seedbed' in which many organizations were able to foster such linkages and utilize them in other areas of their work. In this respect the RTP Steering Group and other interactive events represented fora for setting precedents and building confidence between regional actors. An interesting example of relationship-building and learning between support actors occurred during an RTP seminar held specifically for the region's higher education industrial liaison officers (ILOs). This two-day event represented the first occasion that ILOs from all areas of Wales had been brought together. As such it was widely recognized by participants as a constructive and valuable exercise in terms of learning about each other and helping to encourage stronger bonds within the sector. In particular, the event succeeded in producing a remarkably objective and open discussion about both individual institutions and collective weaknesses in innovation support. It also generated discussions about the possibility of organizing an electronic forum to facilitate networking in the future. More recently the RTP has begun to implement a range of networking programs (set out in the Action Plan) designed to stimulate innovation. It is perhaps here, then, that the exercise has the greatest *potential* for generating new interactions amongst firms and between firms and support institutions.

The third domain in which the Wales RTP has impacted on learning processes has been its role in embedding more inclusive regional policy routines. Indeed, amongst many participants there is widespread agreement that the RTP has represented an important break from the past in terms of strategy-making exercises. Perhaps the most prominent example of this has been the way it has shaped the revision of the most recent EU regional development program for South Wales. This marked an important change because it allowed the program, for the first time, to be developed on the basis of a systematic assessment of needs and consensus. Such was the perceived strength of this process, that the regional Monitoring Committee felt sufficiently confident to increase the amount of funding for the 1997–99 program devoted to innovation measures.

Outside the region's EU development program the impact of the RTP process has also been felt in other strategy-making exercises in the region. Notable, here, has been the 'spill-over' effect on the Wales Information Society[4] program launched in July 1997. This has adopted a similar inclusive and interactive learning approach to strategy-building (Osmond, no date). It has also incorporated several individuals from the RTP's Management Unit and Steering Group in an attempt to transfer some of the skills and capabilities acquired during the process. Elsewhere, the recent RTP consultation exercise has also revealed the desire, on the part of organizations in mid- and north-west Wales, to explore the possibility of establishing subregional strategy exercises to complement the all-Wales RTP. These developments have been acknowledged in RTP 1998, which includes a further pledge to 'work with local authorities, the private sector, Training and Enterprise Councils, the 'economic powerhouse', higher and further education and others to relate the RTP to local needs and circumstances' (WDA, 1998).

It is difficult to say, yet, whether these experiences represent a shift towards the new, inclusive 'culture of strategic decision making' anticipated by the European Commission. Indeed, the recent work of Phelps *et al.* (1998) suggests that in the area of inward investment promotion there is just cause for caution, not least because of the hierarchical and selective nature of strategic collaboration evidenced in the attraction of the LG plant to Wales, the largest inward investment project in Europe. What our research has revealed, though, is a broad desire for the *future* (regional) policy of the Welsh Assembly to adopt a philosophy of inclusive interactive learning.

Before moving on, it is important to try and set the Wales RTP experience in the wider context of RIS exercises undertaken elsewhere in Europe. The first point that should be made here relates to the diversity of regions and institutional terrains in which RIS exercises have been established. Perhaps unsurprisingly, given the bottom-up, locally attuned nature of the exercises, this diversity has produced wide variations in the nature of the processes and subsequent efforts to implement the strategies. Many, for example, were led by institutions with a high degree of regional autonomy. Others with less autonomy, however, relied on leadership from local actors such as universities, development agencies and so on. Likewise, process activities varied across participant regions. This was most evident in the timing and arrangements surrounding the research, consultation and implementation activities in each region. What subsequent research has revealed, however, is that few RTP or RIS regions have adopted the rather linear process anticipated in the Commission's guidelines (Nauwelaers *et al.*, 1996). Indeed, most have been characterized by high levels of overlap and feedback between the various activities. Thus, in Limburg, for example, pilot projects were implemented early on in the process, alongside discussion and research, as a means of demonstrating intent and providing experimental feedback on particular project mechanisms which could later be implemented. Others, like Wales, however, chose to view implementation as the end product of a lengthy series of discussion and consultation exercises. In the light of this diversity it seems clear that we should talk not of a single, best-practice model, but of rich diversity and context specificity.

Fortunately, this insight has been woven into the first major independent evaluation to be undertaken of the RTP and RIS exercise (Arnold *et al.*, 1998). In short, the results of this study suggest that gains can invariably be divided into two main areas (broadly corresponding with our earlier framework used in the Welsh case study): first, the creation of a regional forum that improves informal rules and facilitates consensual policy-making; and second, the adoption and implementation of new forms of needs-based innovation policy. According to Tsipouri (1998) most of the participant regions have managed to achieve the first outcome, while only a handful have been able to succeed in terms of the second. In this respect, then, it is probably best to say that the program has enabled most regions to build capacities in the form of relational assets, knowledge acquisition and new thinking. It is less clear, though, whether other regions have been able to achieve results, in terms of shaping the activities and strategies of support institutions, in the way Wales has been able to.

The findings of the EU evaluation conclude by suggesting that the RIS exercise tends to work best in regions which have a high degree of regional autonomy. This, it is argued, can enable regional administrations (as participants in the exercise) to plan and implement priorities without conflicting demands and barriers from the national level. In the absence of this link between the strategy and policy-making, the exercise has been short-circuited in a number of regions. In Lorraine, for example, Nauwelaers (1999) has described how despite facilitating an intraregional learning process, the strategy largely became subsumed within a broader national policy agenda in a way which diluted both its scope and emphasis.

A more general feature of the RIS planning process was its inability to ensure that poorly performing institutional structures and policy routines were addressed. Instead, in many cases RIS exercises tended to favor current activities, actors and regional priorities (Tsipouri, 1998). In Wales, for example, this was evident in the way that emphasis was given to working with existing resources and actors to understand the needs of firms, while relatively little attention was given to restructuring or terminating poorly performing or unnecessary initiatives. The explanation for this weakness rests with the presence of pre-existing policy communities and institutions and the need for RIS to build inclusive support for the exercise. These factors, together, can create the conditions in which the discussions within a RIS exercise become limited to incremental rather than radical changes in structures and priorities.

A further issue highlighted in the EU evaluation was that gains from the RIS process tended to be easier to achieve (and detect) in those regions which embraced the ethos of experimental learning. That is, those regions which saw the exercise as an opportunity to facilitate new linkages and learning between actors, rather than simply a source of funds for existing activities, were likely to achieve most from their RIS exercises. This is perhaps an obvious condition, but one which nonetheless contributed towards the failure of a number of exercises to either mobilize significant regional support, or produce a workable strategy document at the end of the 18-month period (Tsipouri, 1998). This attests to the fact that economic development problems can often be deeply ingrained, requiring much longer time frames to build sufficient relational assets.

Despite these potential areas of weakness, it seems clear from our own experiences and recent evaluations that RIS holds significant promise for FRS. The outcomes possible through this program, however, are not necessarily those of traditional regional policy; instead they appear to be primarily centered around softer intangibles such as trust, reciprocity and the resulting institutional routines. In addressing these issues further, we now move on to examine the wider implications of these new departures in EU regional policy.

Conclusions and implications

In this chapter we have argued that a radically new kind of regional policy is emerging in the EU in which the accent is on collective learning and institutional innovation rather than upon basic infrastructure provision. We have also argued that these new regional innovation policies signal the most determined effort to date to build social capital, by which we mean a relational infrastructure for collective action predicated on trust, reciprocity and the disposition to collaborate to achieve mutually beneficial ends. In short, this constitutes a form of regional experimentalism which aims to create mechanisms and structures through which local agents can begin to develop new and more purposeful conversations about joint solutions to common problems as a prelude to building more robust and more sophisticated forms of institutionalized voice. Modest as they are, these are nevertheless important steps in institutional capacity building for regions which have hitherto looked almost exclusively for exogenous solutions to their internal problems.

In these small regional experiments we can begin to discern a new and more innovative form of governance, the hallmarks of which are interactive, associational and network-based, rather than hierarchical or market-based, the respective governance modes of dirigisme and neo-liberalism. In other words, regional experimentalism is concerned not with the scale of state intervention but its mode, not the boundary between state and market but the framework for effective interaction. In associational governance structures, therefore, public and private agents are esteemed primarily for their competence in the network rather than their status in the hierarchy or their power in the market. In this context, the debilitating division of labor between conception and execution, between design and delivery, is breaking down as a result of interactive service provision, where services and projects have been jointly defined in an iterative process of producer–user engagement, a process which transcends the impoverished debate as to whether services should be demand-led or supply-driven (Morgan, 1996).

The wider lessons of these regional innovation pilot projects, particularly as regards how to design, deliver, monitor and evaluate, have been compiled in a collective 'memory bank' of good practice in Europe so that regions embarking on the process for the first time can draw on the practical experiences of the pioneers (European Commission, 1997). These lessons can seem prosaic in the extreme when taken in isolation (like who should assume the lead role in project management, which organizations should be represented on the Steering Committee and how should they report to their constituencies to ensure accountability) but, for regions which have never embraced an associational governance structure, which requires an interactive and consultative policy-making process, these lessons can be invaluable sources of knowledge that simply do not exist in the regional milieu.

Although regional experiments like the RTP and the RIS have triggered some encouraging institutional innovations in their targeted regions the key question which needs to be posed is what constitutes success? To answer this seemingly simple question we need to remember that the ultimate aim of these new regional policies is to raise the innovative capacity of the FRS and this is necessarily a long-term process. Having said that, however, there is one short-term indicator which can be used to monitor the value of these regional experiments and that is the extent to which the key participants, especially the firms, remain involved in a process which they find rewarding even when they are unable to quantify the benefits of involvement because of the intangible nature of these benefits. This is a useful political indicator of success inasmuch as it signals the collective commitment of the regional stakeholders to an important precedent, namely that the trajectory of development can be shaped through purposeful conversations within the region. On its own, however, this indicator could conceal as much as it reveals if the conversations are a façade for an incestuous, self-referential club of regional notables, in other words not a new beginning but 'an aimless huddling of elites' (Sabel, 1995).

To overcome this danger we need a series of other indicators to assess longer-term changes in innovation capacity. Here we can usefully distinguish between linear indicators, which focus on 'hard' outputs (like patents filed, R&D expenditure, level of new product development, workforce qualifications and the like) and interactive indicators (which aim to measure 'soft' processes, like institutional linkages, network formation and information flows and so on). The latter can capture important changes in a region's institutional architecture which are beyond the grasp of more conventional linear indicators (Nauwelaers and Reid, 1995). The first evaluation of the RTP exercise suggests that certain regions, Limburg, Wales and Central Macedonia in particular, have achieved some real success with respect to interactive indicators, though it is too soon to expect much progress on the linear indicators (Arnold *et al.*, 1998).

Another danger is that these new regional policies (which aim to raise innovation capacity) might be judged by the standards of the old regional policies (short-term job creation) and get prematurely jettisoned because they fail to meet these standards. To avoid this danger we need to be clear that more innovation does not necessarily mean more jobs; indeed, some forms of innovation, like process innovation for example, can be job-displacing. The point to establish here is that innovation-oriented regional policy needs to be flanked by, and complemented with, job-creating regional policy because FRS need to become more innovative and more cohesive (Morgan and Nauwelaers, 1999). The FRS cannot regenerate themselves solely through their own efforts, hence regionalists need to appreciate the continuing significance of the national state, particularly in the field of cohesion. This point was underlined by the European Commission when, referring to the role of member-state policies, it said that 'solidarity in the Union begins at home' (European Commission, 1996).

Regional innovation policy, as expressed in the RTP and RIS programs, clearly has its limits. At present these programs are small-scale, low-budget experiments which have yet to be fully mainstreamed even in the regions which have pioneered them. To be effective, therefore, such programs need to be taken up and extended by national and supranational authorities in the EU, otherwise they might atrophy for lack of scale and resources. For all that, regional experimentalism might have some lessons for the 'higher' echelons of the state, particularly as regards governance structures and policy-making processes, where politicians and officials too often think of themselves as tutors rather than learners.

In the UK, for example, one of the government's top civil servants has launched an unprecedented critique of the policy-making process, calling for structures and processes which are uncannily like regional experimentalism. Among other things, he argued that 'policy-making was too hierarchical, too adversarial and insufficiently creative' in large part because of the debilitating division between the design and delivery of policy and because officials, too keen to defend turf, often failed to seek knowledge and ideas from outside government (Timmins, 1998).

The European Commission itself is not exempt from such criticisms and it, too, could benefit from some 'reverse learning' by emulating the interactive governance structures and processes pioneered by the regional experiments it has actually sponsored. In the face of the most momentous changes in its history – such as the deepening of the Union through economic and monetary union and the widening of the Union through enlargement – the European Commission will have to give greater weight to the principle of subsidiarity, which aims to devolve decision-making to the lowest practical level. The effectiveness of this multilayer governance system will depend in no small way on each level – regional, national and supranational – respecting the competence of the others and recognizing their systemic interdependence. In other words, the successful evolution of this system will depend on how it resolves the tension between the traditional conventions of hierarchy and the new associational practices of regional experimentalism.

Notes

1   A good illustration of this point comes from Naples, the 'unemployment capital' of the Mezzogiorno, where Antonio Bassolino won the election for mayor on a program which called for 'a normal and legal city'.
2   In areas such as inter-regional cooperation, technology for regional development, regional planning and urban policy.
3   The concept of 'Flagship' projects was one of the suggestions made by the international experts panel.
4   Part of another EU program (Regional Information Society Initiatives) funded through Article 10.

References

Amin, A. (1999) 'An Institutional Perspective on Regional Economic Development', *International Journal for Urban and Regional Research*, Vol. 23, No. 2, June, pp. 365–78.
Amin, A. and N. Thrift (1995) 'Institutional Issues for the European Regions: from Markets and Plans to Socio-Economics and Powers of Association', *Economy and Society*, vol. 24 (1), pp. 41–66.
Arnold, E., P. Boekholt and L. Tsipouri (1998) 'Evaluation of RTP Pilot Programs', Draft Report to European Commission – DGXIII/XVI.
Axelrod, R. (1984) *The Evolution of Cooperation*, London: Penguin.

Cooke, P. (1998) 'Global Clustering and Regional Innovation: Systemic Integration in Wales', in H.-J. Braczyk, P. Cooke and M. Heidenreich (eds) *Regional Innovation Systems*, London: UCL Press, pp. 245–62.

Cooke, P. and K. Morgan (1998) *The Associational Economy: Firms, Regions and Innovation*, Oxford: Oxford University Press.

European Commission (no date) *Proposal of a Pilot Action to Produce Regional Technology Strategies in Objective 1 and 2 Regions and the New German Lander*, Brussels: Commission of the European Communities.

European Commission (1993) *Growth, Competitiveness and Employment*, Brussels: Commission of the European Communities.

—— (1994) *Regional Technology Plan Guide Book* (2nd Edition), Brussels: Commission of the European Communities.

—— (1995) *Green Paper on Innovation*, Brussels: Commission of the European Communities.

—— (1996) *First Report on Economic and Social Cohesion*, Brussels: Commission of the European Communities.

—— (1997) *Practical Guide to Regional Innovation Actions*, Brussels: Commission of the European Communities.

—— (1998) *Reinforcing Cohesion and Competitiveness through RTD and Innovation Policies*, COM (98)275, Brussels: Commission of the European Communities.

Henderson, D. and M. Thomas (1999) 'Learning through Strategy-making: the RTP in Wales', in K. Morgan and C. Nauwelaers (eds) *Regional Innovation Strategies: the Challenge for Less Favoured Regions*, London: Jessica Kingsley.

Hirschman, A. (1958) *The Strategy of Economic Development*, Yale: Yale University Press.

—— (1970) *Exit, Voice and Loyalty: responses to decline in firms, organizations, and states*, Cambridge: Harvard University Press.

—— (1986) *Rival Views of Market Society and Other Recent Essays*, New York: Viking.

Lampedusa, G. (1960) *The Leopard*, London: Harvill.

Landabaso, M. (1997) 'The Promotion of Innovation in Regional Policy: proposals for a regional innovation strategy', *Entrepreneurship and Regional Development*, vol. 9, pp. 1–24.

Landabaso, M. and A. Reid (1999) 'Developing Regional Innovation Strategies: the European Commission as Animateur', in K. Morgan and C. Nauwelaers (eds) *Regional Innovation Strategies: the Challenge for Less Favoured Regions*, London: Jessica Kingsley, The Stationery Office.

Luhmann, N. (1979) *Trust and Power*, H. Davis, J. Raffan and K. Rooney (trans), Tom Burns and Gianfranco Poggi (eds), Chichester: Wiley.

Maskell, P. *et al.* (1998) *Competitiveness, Localised Learning and Regional Development: specialisation and prosperity in small open economies*, London: Routledge.

Messina, C. (1997) 'Innovative Actions under Article 10 of the ERDF: Editorial', *Newsletter of Pilot Projects*, 1, pp. 9–97.

Morgan, K. (1996) 'Learning by Interacting: Inter-Firm Networks and Enterprise Support', in OECD (ed.) *Networks of Enterprises and Local Development: competing and cooperating in local productive systems*, Paris: OECD.

—— (1997) 'The Learning Region: Institutions, Innovation and Regional Renewal', *Regional Studies*, vol. 3 (5), pp. 491–503.

—— (1997) 'The Regional Animateur: Taking Stock of the Welsh Development Agency', *Regional & Federal Studies*, vol. 7 (2), pp. 70–94.

Morgan, K. and Nauwelaers, C. (eds) (1999) *Regional Innovation Strategies: the Challenge for Less Favoured Regions*, London: The Stationery Office.

Nauwelaers, C. (1999) 'RTP in Lorraine', in K. Morgan and C. Nauwelaers (eds) *Regional Innovation Strategies: the Challenge for Less Favoured Regions*, London: The Stationery Office.

Nauwelaers, C. and A. Reid (1995) *Innovative Regions*, Brussels: European Commission.

Nauwelaers, C., J. Cobbenhagen, J.C. Moretti, J. Severijns and M. Thomas (1996) 'Building Regional Innovation Strategies: RTPs in an Evolutionary Perspective', *Maastricht Economic Research on Innovation and Technology Working Paper 2/96–017*.

Osmond, J. (no date) *Wales Information Society*, Cardiff: IWA/WDA.

Phelps, N. A., J. Lovering and K. Morgan (1998) 'Tying the Firm to the Region or Tying the Region to the Firm?', *European Urban and Regional Studies*, vol. 5 (2), pp. 119–37.

Piore, M. (1995) *Beyond Individualism*, Cambridge: Harvard University Press.

Powell, W. (1996) 'Trust-Based Forms of Governance', in R. Kramer and T. Tyler (eds) *Trust in Organizations: frontiers of theory and research*, Thousand Oaks, CA: Sage, pp. 51–67.

Putnam, R. (1993) *Making Democracy Work: civic traditions in modern Italy*, with Robert Leonardi and Raffaella Y. Nanetti, Princeton: Princeton University Press.

Sabel, C. (1994) 'Learning by Monitoring: the Institutions of Economic Development', in N. Smelser and R. Swedberg (eds) *Handbook of Economic Sociology*, Princeton: Princeton University Press, pp. 137–65.

—— (1995) *Experimental Regionalism and the Dilemmas of Regional Economic Policy in Europe*, Paris: OECD.

Storper, M. (1997) *The Regional World: territorial development in a global economy*, London: The Guilford Press.

Timmins, N. (1998) 'Top Mandarin Takes Government to Task Over Making of Policies', *Financial Times*, 14 December, London.

Tsipouri, L. (1998) 'Lessons Drawn from the RTP/RITTS/RIS Projects', Draft Report to DGXVIII/GXVI, Brussels.

Welsh Development Agency (1996a) *Wales Regional Technology Plan: Consultative Report*, Cardiff: WDA.

Welsh Development Agency (1996b) *Wales Regional Technology Plan: Action Plan*, Cardiff: WDA.

Welsh Development Agency (1998) *Wales Regional Technology Plan: Review and Update, 1998*, Cardiff: WDA.

Williams, J. (1980) 'The Coalowners', in D. Smith (ed.) *A People and a Proletariat: essays in the history of Wales 1780–1980*, London: Pluto Press.

11

Negotiating Order: Sectoral Policies and Social Learning in Ontario[1]

David A. Wolfe

Introduction

Innovation and learning are closely linked in the literature on technological change and the global economy. From the perspective of both evolutionary economics and regional studies, the capacity to innovate in turn is linked to the ability to harness successfully new knowledge in the pursuit of commercial applications or fashion significant improvements in product and process technologies. This trend has led many observers to describe the industrial countries as knowledge-based economies, however, it may be more appropriate to describe them as a learning economy. As the rapid pace of change associated with the 'frontiers' of economically relevant knowledge accelerates, the economic value of individual pieces of knowledge diminishes the more widely accessible they become. The increased availability of information resulting from the rapid diffusion of new information technologies and the emphasis on learning are thus linked. Learning in this respect refers to the building of new competencies and the acquisition of new skills, not just the acquisition of information (Lundvall and Borrás, 1998, p. 35).

The capability of firms, regions and nations to learn and adapt to rapidly changing economic circumstances will most likely determine their future success in the global economy. Learning as a part of the innovation process is strongly conditioned by the broader social and cultural context in which it is embedded. The firm's capacity to absorb new knowledge and improve its internal routines for learning is strongly conditioned by the broader societal factors that shape its environment. Given that learning, as opposed to the dissemination of knowledge, is a highly localized process, regions are coming to be seen as a relevant site or level of study for those trying to analyze or understand the factors that contribute to, or impede, the potential of organizations to develop new learning capacities.

These facts pose a particular challenge for older industrial regions with mature or established economies, such as those in the industrial heartland of North America and Europe. In these economies, institutional practices are embedded in well-established cultural and social practices. In some instances, these practices may support innovation and social learning, but in others, they may not be particularly well suited to the institutional requirements of the learning economy. In these cases, the need to 'forget' may be a prior condition of the ability to learn. The inertial effect exerted by the power of old routines and habits may block the ability of firms or networks to develop new learning processes (Gregersen and Johnson, 1997, p. 480). The challenge is particularly great for those economies that are based in a market-oriented, liberal politics with a weak history of collaborative relations and networking, either between firms or between the public and private sectors. Societies with a weaker tradition in this regard may experience special challenges in making the transition to a more innovative and learning economy.

The current case study, based on Ontario, the largest province and industrial heartland of Canada, represents just this kind of instance. The Ontario economy is dominated by more traditional manufacturing industries, especially the auto industry, and its industrial culture is characterized by features associated with the liberal or Anglo-American business model, which is governed by the values of rugged individualism, self-sufficiency and competitive rivalry. The election of a New Democratic Party government in 1990 marked a significant turning point in the history of the province. It represented a political breakthrough for Canada's social democratic party in the country's most populous province, but unfortunately, it came just as the economy entered its most severe recession since the 1930s. The impact of the cyclical downturn was compounded by a process of structural transformation triggered by the broader forces of technological change and Canada's integration into the North American Free Trade Agreement. The task of crafting a distinctive approach to industrial policy and structural adjustment was one of the most challenging faced by the NDP government in its attempt to promote the conditions conducive to a learning economy.[2] This chapter explores one of the centerpieces of this approach – the formation of distinctive sector strategies in an attempt to foster a process of social learning across broad sectors of the provincial economy.

Social learning and institutional reflexivity

The challenge of effecting a change to a learning economy raises important questions about the appropriate role for the state and public policy. Some have suggested that the state, especially at the regional or subnational level, retains a positive role to play in the transition to a learning economy, but it requires a different conception of the state from that which has traditionally prevailed. The kind of interactive learning described in the introduction to this volume requires a degree of *reflexivity*, or the ability to self-monitor and learn from past successes and failures, in other words, to learn how to learn. This notion of institutional *reflexivity* poses important questions about the nature of public policy formation in modern democracies; for it is not only private institutions that must learn and adapt to the changing realities of a more innovative economy, but public ones as well. In the traditional Weberian conception of bureaucracy, the administrative apparatus is the repository of all relevant policy knowledge and expertise. The policy revolution of the postwar period, with its emphasis on the rational approach to policy formation, reinforced this traditional foundation. This approach downplayed questions of conflicting interests in the formation of policy and highlighted the potential contribution that expert analysis could make to solving complex problems.

Yet the more recent acceptance in the policy literature that rationality is bounded and that conflicting choices and values underlie our conception of the public interest leads to a more contingent approach. In this alternative perspective, the role of policy analysis contributes to the discourse and bargaining within which public policy is formed. The organizational and institutional structures within which policy is formed are also critical. The design of appropriate policy depends to a large extent on the design of organizational structures capable of learning and adapting to what is learned. This concept is not entirely novel; it has begun to appear with increasing frequency in the literature on 'policy-making as social learning'. Yet, as Hall points out, the concept has been presented in only the sketchiest of terms and for the most part, it has been rooted in the literature that emphasizes a strong degree of state autonomy. To the extent that social learning describes the policy process, it applies to a process internal to the state (1993, pp. 275–6). As such, it remains grounded within the traditional Weberian conception of the state and its bureaucracy.

The idea that institutional learning is relevant to policy-making must go beyond the stage of policy formation. Although policy is formulated at the highest levels of government, it is generally carried out by lower levels of government, frequently in interaction with private parties. The actual policy as it is implemented on the ground involves working out conflicts among winners and losers as it is implemented; the broadly defined policy may change in the process. What ends up being implemented often differs radically from what the policy-makers originally had in mind (Majone and Wildavsky, 1984). The effectiveness with which policies are implemented depends on the capacity of the institutional structures to adapt to this reality.

> . . . if one views policy making as a continuing process, the organizational and institutional structures involved become critical. Public policies and programs, like private activities, are embedded in and carried out by organizations. And in a basic sense, it is the organizations that learn, and adapt. The design of a good policy is, to a considerable extent, the design of an organizational structure capable of learning and of adjusting behaviour in response to what is learned.

. . . just as many analyses of the workings of the market economy tend to abstract the private economy from public policies, programs and institutions, too many analyses of public policies and programs do not recognize adequately that their effects will be determined, to a considerable degree, by private and not governmental actors (Nelson and Winter, 1982, p. 384).

This insight offered by Nelson and Winter suggests a radically different approach to public policy formation from that conventionally followed. It substitutes the exclusive role of the public bureaucracy for a more mixed public/private model and emphasizes the context of institutional structures and learning. It involves the devolution of more autonomy and responsibility for the policy outcome onto those in the private sector directly affected. It corresponds closely with the concept of the associational state that has emerged recently.

In a number of articles, including his contribution to this volume with Dylan Henderson, Kevin Morgan outlines a conception of the associational state more suitable to the context of a learning economy. In his view, the key issue is not the scale of state intervention in the economy, but rather its mode of intervention. The key factor is not the boundary drawn between the state and private economic actors, but rather a framework for appreciating the effective interaction between the two. One of the key challenges for the state is to create the conditions in which firms, associations, and public agencies engage in a collective process of interactive learning that is essential to innovation in the modern knowledge-based economy (Morgan, 1999).

Amin suggests the associational model embodies a conception of the *reflexive* state that includes four key principles. The first is a degree of decision-making pluralism, which involves delegating decision-making authority to the levels and bodies at which policy effectiveness can best be achieved. The second involves the notion that the state provides strategic leadership and the capacity to coordinate. This is not a role that follows from the politics of command and control. Effective leadership requires the combining of authority with a capacity for consensus-building in the appropriate arenas. The third point involves the adoption of a process of dialogic rationality. The relevance of dialogic democracy involves a lasting consensus that results from interactive reasoning. The fourth point involves the commitment in the process of democratic practices to transparent and open government.

The appeal of the associative model of governance lies precisely in the fact that it substitutes a mix of public and private roles for the exclusive role of the public bureaucracy. It devolves a greater degree of responsibility for the policy outcome onto those organizations that will either enjoy the fruits of the policy success or live with the consequences of its failure. Wolfgang Streeck and Philippe Schmitter describe a similar model of associative order as a form of 'private interest government'. They employ the term to refer to the self-government of specific categories of social actors based on their collective self-interest. They also restrict the use of the concept to arrangements where efforts are made to make associative, collective action contribute to the achievement of public policy goals. For Streeck and Schmitter, 'the corporatist-associative delegation of public policy functions to private interest governments represents an attempt to utilize the *collective self-interest* of social groups to create and maintain a generally acceptable social order, and it is based on assumptions about the behavior of *organizations* as transforming agents of individual interests' (Streeck and Schmitter 1985, p. 129).

Despite assertions to the contrary, the adoption of an associative model does not necessarily imply an abandonment of a central role for the state, but rather a rethinking of its role. In an associationist model, the relevant level of the state is to be one of the institutions of the collective order, working in relationship with other organizations, rather than operating in its traditional command and control fashion. Streeck and Schmitter refer to their notion of 'private interest government' as a mixed mode of policy-making where the associative order emerges through the form of a mixed politics (Streeck, *et al.*, 1985, p. 134). The state in this model continues to establish the basic rules governing the operation of the economy, but it places much greater emphasis on the devolution of responsibility to a wide range of associative partners through the mechanisms of 'voice' and consultation (Amin, 1996, p. 19). The key challenge in the associationist model is to achieve the most effective balance between the state's need to provide direction and the desirability of providing greater 'voice' through the devolution of responsibility.

The associational conception of the state also implies the devolution of power in the state system from remote bureaucratic ministries at the national level to local and regional levels of government better positioned to build lasting and interactive relations with firms and business associations in their regions. In addition, it may involve the delegation of certain tasks like enterprise support services from formal government agencies to accredited business associations because the latter possess relevant assets, such as a knowledge of, and credibility with, their members, which the state needs to enlist in order to ensure the effectiveness of its support policies. Devolving power to the lower levels of government creates the opportunity for more meaningful dialogue to take place at the regional level. This is important because dialogue or discussion is central to the process by which parties come to reinterpret themselves and their relationship to other relevant actors within the local economy (Morgan, 1999).

Building trust among economic actors in a local or regional economy is a difficult process that requires a constant dialogue between the relevant parties so that interests and perceptions can be better brought into alignment. Authors, such as Charles Sabel (1992) and Michael Storper (this volume) underscore the critical role played by soft factors, such as talk, in building

trust and the kind of long-term relationships that underpin the institutionalized learning economy. Storper suggests that talk and confidence are more
likely to succeed when they occur in a setting that is geographically localized
and that small, repeated low-cost experiments can help to generate interactive learning between parties in an environment which has previously
been characterized by distrust or antipathy. Morgan notes that these same
concerns lie at the heart of the regional innovation experiments currently
underway in Europe. To those more familiar with the economic environment and business culture of North America, this characterization seems
somewhat optimistic to say the least. The critical issue is whether it is possible to create the necessary conditions of trust and interaction required to
achieve this level of cooperation.

The literature on reflexivity and associative governance prescribes a potential path that governments can take in attempting to foster the conversion of
their respective economies to that of the learning economy, but it does not
offer much help on how or why the institutional structures of different
regions and nations may or may not support this capacity for institutional
learning. The comparative politics literature affords a different perspective
on this question. The institutional structures within which policy is formulated and implemented reflect the outcomes of past struggles and alliances
among social and political actors as refracted through their economic interests – primarily, but not necessarily, their interests as producers. Shifts in the
resulting policy regimes can and do occur as a consequence of shifts in
the perceived interests or power relations among the relevant actors. Therefore
it is essential to specify the historical context within which the policy regimes
are formulated (Pontusson, 1995, pp. 140–1; Gourevitch 1986).

The broader strategies and goals of public policy discussed above depend
to a great extent on the organization of industry within a specific territorial
unit and the sectors within that industry. To the degree that industry already
enjoys a cohesive organizational culture or has a strong set of industry or sector-specific associations with a tradition of acting collectively to solve its
problems, there will be a stronger basis for the industry or sector to search
collectively for new solutions to economic challenges or deal with the problems of structural adjustment. However, to the extent that the sector is
characterized by a more fragmented and competitive business culture, the solutions chosen will most likely reflect this underlying culture. Hollingsworth and
Streeck suggest that the relations among firms within a sector are shaped by
the distinctive properties of the sector – especially by the properties of technology on the one hand and products and product markets on the other – as
well as by broader national or regional differences in culture and institutions. Patterns of relations with individual sectors and countries evolve over
time and determine the social conditions under which firms must adapt
to broader changes in technology and the economy (Hollingsworth and
Streeck, 1994, p. 278).

The comparative study of how different countries and regions respond to the challenge of economic change and structural adjustment suggests the possibility of a variety of possible responses. One which is relevant to the current case, is the response by Scottish industry to the problems of adjustment faced in the 1980s. In their study of this process, Chris Moore and Simon Booth propose a simplified typology of possibilities. Between the more conventional responses of macro-level corporatism and the traditional forms of pluralism, they suggest a third possibility for arriving at sectoral level solutions, which they term a negotiated order. The arrangements derived from this form of negotiated order combine some degree of public and private interests at the sectoral level in the formulation and implementation of an agreed strategy. They differ both from the traditional market-led approach to adjustment to the extent that private associations and interests assume some degree of collective responsibility for the agreed-upon policy. But arrangements of this nature lack the consistency and coherence usually associated with more formal corporatist arrangements because they lack the hierarchical order and the ability to discipline their members that characterize those types of arrangements:

> These arrangements inhabit a world between corporatism, market and the State. In these sectors both government and private interests recognize the benefits of consensus in order to achieve either a greater strategic advantage for individual companies or to aid economic adjustment (Moore and Booth, 1989, p. 84).

Although Moore and Booth do not refer to the concept of associative governance, their definition of a negotiated order bears a strong affinity to the type of intermediary relations between the state and private economic actors discussed above. The negotiated order involves a higher degree of consensus around an issue than is normally present in the pluralist type of arrangements found in most industry relationships with government, but lacks the institutional resources and the effective power to transform the agreed-upon strategy into policy. Thus it still requires a certain degree of leadership by political actors to give form and content to the consensus forged within the sectoral order. One of the advantages of the negotiated order over more formal corporatist arrangements is that neither the formal political authorities nor sectoral groups need surrender any of their autonomy to the other – both the public and private sector participants can claim to have preserved their independence (Moore and Booth 1989, p. 116). For the Government of Ontario, trying to cope with the economic challenge facing the province in the early 1990s, the prospect of establishing a negotiated order in some economic sectors contained the potential for creating a more associational form of governance that could underpin the transition to a learning economy.

The sector strategies in Ontario's economic development policy

The critical policy challenge faced by the NDP government in 1990 lay in responding to the major structural changes affecting the provincial economy resulting from the trend towards continental integration and the impact of technological change. The problem was compounded by the relative underdevelopment of industrial policy at the provincial level.[3] Furthermore, the NDP government inherited a set of public institutions with little responsibility for, or experience in, dealing with the issue of industrial adjustment. However, the election of the NDP signaled an important shift in the institutional structures within which policy is formulated and implemented. The election result signified such a shift in three important respects. In the first instance, it fundamentally transformed the traditional relations between the business community and the government in Ontario. Traditional avenues of access were closed and to some extent a more level playing field was created for different segments of the business class. In the second place, organized labor was clearly afforded a more equal place at the table than they had enjoyed for many years. While they exerted little direct influence on the formation of industrial policy, their inclusion represented an important change in the social base of support for the new policy regime. Finally, the election of the NDP also sent a positive signal to those elements of the public bureaucracy interested in a more activist approach to industrial policy.

However, the NDP itself was not particularly well prepared for the task of framing a policy response suitable for the economic situation they faced. In part, this had much to do with the particular style of opposition politics that it had refined with great success during its years in the political wilderness (Rachlis and Wolfe, 1997); in part, it grew out of a policy discourse that was trapped in the economic realities of the 1970s more than the 1990s. An early attempt to lay out the framework for a new economic approach was presented in the background paper released with the April 1991 budget, *Ontario in the 1990s*. The paper argued that the goal of sustainable prosperity could best be realized through the creation of high value-added, high-wage jobs, and strategic partnerships. The major challenge in the 1990s was to increase the overall productivity of Ontario's economy, not by minimizing cost levels for the existing mix of product and processes, but by promoting continuous improvement in products and processes across the networks of firms and sectors in the provincial economy (Laughren, 1991).

The budget paper raised several issues related to the concept of associative governance through a discussion of the role of partnerships. The government indicated that the realization of its economic strategy must be based on a broad social partnership. It required strategic public and private initiatives in a climate that allowed the respective partners to develop a sense of collective responsibility. A concerted and cooperative approach was deemed essential to achieve the government's goals with respect to economic development and sustainable prosperity (Laughren, 1991, p. 101). The emphasis placed on the role of partnerships indicated that the government was prepared to contemplate changes in the way that it worked with the private sector to deliver economic development policies. It also suggested that the government was attempting to forge a new analytic and policy framework for the delivering of its economic development policy – one which diverged from the traditional command and control form of policy implementation and from the use of traditional policy instruments such as tax incentives and direct expenditures to accomplish its goals.

The direction of the government's economic development strategy was spelled out more clearly in the *Industrial Policy Framework*, released in July 1992. The framework reiterated the overall goal of facilitating the transition of the Ontario economy towards those sectors and firms with the capacity to generate higher wage, higher value-added, and environmentally sustainable jobs. It focused on ways of developing higher value-added activities throughout the economy to increase competitiveness and create more, and better, jobs (Ontario, Ministry of Industry, Trade, and Technology, 1992). One of the key changes envisioned in the framework was the increased emphasis on working with sectors. The focus on sectors was not an entirely novel departure either for the Government of Ontario or the federal government. The Ministry of Industry, Trade and Technology had a tradition dating back to the early 1980s of working with individual sectors to strengthen their competitive position and promote their sectoral capabilities. Notable examples of this activity included efforts to strengthen the position of Ontario's film and television industry, the radical restructuring of the wine industry under the Ontario Winery Adjustment Program and the food processing development strategy previously recommended by the Food Industry Advisory Committee. In addition, MITT had developed an internal sectoral capacity through its dealing with individual sectors, such as the automotive industry and the furniture industry. The sectoral development process was designed to build upon existing capabilities within the public sector to the greatest extent possible to enhance the strong working relations that already existed between individual ministries or branches and specific sectors. In instances where these relations did not exist, ministry leads were assigned as prospective sectors identified themselves.

The Sector Partnership Fund (SPF) announced in the budget of April, 1992, was a three-year initiative (later extended to six), budgeted at $150 million, and designed to implement the sectoral component of the Industrial Policy Framework. The Sector Partnership Fund was designed to provide assistance to cooperative sector projects based on an approved strategy fashioned by the key players within the sector. Its overall objectives were to improve the competitiveness of sectors and foster their development by promoting the shift to higher value-added activities. For the purposes of the sector development process, a sector was defined as a group of Ontario-based firms that produce similar goods and services, that identified themselves as a sector, had a recognized association or forum for resolving sector-specific issues, had identified a range of sectoral issues of concern to a broad cross-section of members, and had multipartite representation, including business, labor, and other relevant stakeholders in the sector.

Funding under the Sector Partnership Fund was based on the four principles of flexibility, cooperation, leverage and accessibility. It recognized that each sector faced unique competitive challenges and was designed to be flexible in responding to those circumstances. Individual industrial sectors are characterized by distinctive sectoral properties, shaped by the specific nature of the technology they use and the constraining effects of their products and product markets. A critical principle incorporated into the SPF was that of leverage. In a time of scarce fiscal resources, the government maintained that it could not, and should not, assume full responsibility for funding sector-based initiatives. Sector Partnership Fund support was intended to lever project funding from industry, labor and other levels of government. The process was based, in part, on the assumption that eligible projects constituted a form of quasi-public goods, whose utility to industry partners was strong enough to attract some private investment, but insufficient to be self-financing. It was also seen as a way of subjecting the sectoral initiatives to a form of market test – to determine if the private sector was willing to support them itself. Finally, the principle of accessibility established that all sectors were deemed potentially eligible for funding and that within each sector a substantial proportion of its participants must stand to benefit from SPF-supported initiatives.

To qualify for funding under the Sector Partnership Fund, a sector had to meet a number of rigidly specified criteria. It had to fashion an approved sector strategy developed through a broadly based multipartite consultation process. While not explicitly corporatist in its approach, the Sector Partnership Fund required that all labor market partners in the sector be involved. Industry associations, labor organizations and other stakeholders were encouraged to consult together in the identification of common challenges facing the sector and the development of sectoral strategies for submission to the government. Funding up to a limit of $500,000 was available from the SPF to support the formation of an approved strategy, although the approval process to access these funds was also quite stringent. The sector strategies were expected to address a common range of issues, including: a comprehensive review of its external environment addressing both the threats and opportunities that it expected to face in the medium and long term; a comprehensive analysis of its internal strengths and weaknesses and how they affected its ability to respond to the external environment; an attempt to forecast several future scenarios for the sector based on its assessment of both the external environment and its internal status; and finally, a strategic plan developed out of the preceding analysis that established objectives for the sector to improve its competitiveness, move to higher value-added activities, and lay out an agreed-upon strategy to achieve those objectives. The plan could then identify specific initiatives to be funded from the SPF that flowed from its strategic plan.

Once completed, the sector strategies were submitted to the Cabinet Committee on Economic Development for approval. Following this stage, specific initiatives could be submitted to the government by the sector and its sponsoring ministry for funding. The specific initiatives had to address common challenges or needs within the sector that had been identified in the sector strategy; they had to ensure that assistance was accessible to a wide range of participants within the sector; the assistance provided was to be incremental to funding already available to the sector; and finally, the initiatives undertaken were not to substitute for activities that would have been undertaken without SPF participation. Both the approval process itself and the criteria for accessing funding were extremely rigid and bureaucratically cumbersome, reflecting the concern by some members of the government that this not become another means for subsidizing individual firms. However, the rigid nature of the criteria became a point of contention between several of the sectors and government officials.

Five specific types of initiatives were deemed eligible for funding under the SPF: 1) developing sectoral technological capability; 2) sharing sector knowledge and know-how; 3) sector promotion and marketing (including exports); 4) creating specialized infrastructure; and 5) focused upgrading programs. Developing sector technological capabilities meant promoting the diffusion of advanced technology throughout the sector to better manage their own R&D efforts, improve their production processes, or develop new products and services. Sector promotion and marketing meant helping Ontario-based firms establish new markets or more fully exploit their opportunities in current ones. Sharing sector knowledge included efforts to facilitate the flow of information through related firms within a sector. Creating specialized infrastructure recognized the important contribution that sector-specific institutions could make to the competitiveness of individual firms. Instances of this type of infrastructure could include the computerization of supplier networks within a sector, the establishment of sector-specific training institutes and efforts to strengthen the role of sectoral associations. Finally, the SPF allowed that in unique circumstances, the competitive basis of the sector could be best enhanced at the level of the individual firm through measures such as diagnostic audits of the firms' technical capabilities or programs to assist a wide range of firms to upgrade those capabilities.

The sector strategies in operation

The boldness of the sector strategy process lay, in part, with the range of sectors that the government successfully involved in the process, as well as in the extent to which it devolved responsibility for the content of the strategies onto the sector partners. This grew out of the belief that governments had to break with the hierarchical mode of policy formation and implementation and, in the words of Nelson and Winter, recognize that the outcome of 'their effects will be determined, to a considerable degree, by private and not governmental actors'. Despite the previous experience of both the Ontario and federal governments in working with sectors, the extent of the challenge involved was considerable. The objective of the approach ran counter to the prevailing Anglo-American business culture in Ontario, which is dominated by the ideal of rugged individualism, self-sufficiency and competitive rivalry. In Coleman's study of business associations, he noted that 'business leaders in Canada possess what might be called an individualistic industry culture' (Coleman, 1988, p. 5). This industry culture is reflected in the attitudes and practices of individual firms, whose management remain suspicious and skeptical of collective action and cooperation. The process of interfirm cooperation in Ontario's manufacturing industries is discouraged by the absence of the strong sectoral coordinating mechanisms among firms provided by industry associations or Chambers of Commerce in some of the more innovative regional economies in Europe.

Furthermore, the majority of industry associations in Canada are organized on a national, rather than a regional or provincial basis (with the exception of Quebec), and subprovincial organizations are even more rare. This weakness is compounded by the virtual absence of provincial level sections of any of the major, sector-spanning associations that operate at the national level. The business associations that do exist are fragmented for the most part and tend to represent specialized product interests. There are few peak associations that aim to aggregate the views and interests of these more narrow and sectoral associations. Furthermore, the comprehensive business associations that do exist all enroll firms directly as members. 'Each is a champion of the autonomy of the individual firm and each treats with suspicion the idea of concertation with state officials on economic policy' (Atkinson and Coleman, 1989, p. 48).

The adoption of the sector strategy approach and the creation of the Sector Partnership Fund by the NDP government were an explicit attempt to alter this aspect of the business culture. In effect, the strategy was an attempt to influence the business culture of the province in the direction of creating *socially* organized, firm-based systems for learning, collaboration, cooperation and regulation, initially through the mechanism of encouraging sector participants to *talk*. As was noted at the outset, this kind of facilitated interaction can play a valuable role in building trust and the kind of long-term relationships that underpin the institutionalized learning economy. By any criteria of measurement, the initial stage of sector consultation and strategy formation must be viewed as a success.

Both the number of sectors involved and the extent of participation by key sector players vastly exceeded the initial expectations of the government. Between the summer of 1992 and the election in June 1995, the lead ministries in the government worked with a wide range of stakeholders to develop sector strategies. Consultative efforts produced approved strategies in 15 sectors: Food Processing, Green Industries, Telecommunications, Computing, Tourism, Cultural Industries, Aerospace, Auto Parts, Mines and Minerals, Construction, Health Industries, Forestry, Plastics, Residential Furniture, and Chemicals. By the spring of 1995, work plans were approved and strategies under development in a range of additional sectors, including: Biotechnology, Consulting Engineering, Design, Machinery, Tool, Die and Mold, Retail, and the Electrical and Electronics industry. The last of these strategies was finally released in May 1996. In each case, the consultative efforts drew in as many as 150 individuals in the sector to prepare detailed analyses of sectoral strengths and weaknesses and propose a course of action. In the end, the approach involved 28 different sectors and over two thousand individual participants, representing 22 different unions, 93 industry associations, and 28 universities and colleges (Ontario, Ministry of Economic Development and Trade, 1995).

The following discussion presents an analysis of how the sector development process worked in a number of key sectors and an assessment of whether it realized the ambitious objectives set for it. The assessment is based on a detailed survey of nine of the sectors for which full-fledged strategies were developed: Telecommunications, Computing, Culture, Health Industries, Auto Parts, Aerospace, Plastics, Electrical and Electronic, and Machinery, Tool, Die and Mold. In each case, a set of detailed interviews were conducted with between 10 and 20 participants in the strategy process and the strategies were assessed in terms of the strategic plans formulated and the concrete initiatives arising from the strategy that were approved.[4] Assessing the outcome of the process is complicated somewhat by the fact that the strategies reflected the principles on which they were constructed, namely, that of flexibility. As a result, each strategy was unique, which adds to the richness of the analysis that follows.

The reasons provided by sector participants for participating were quite varied. For many in the business community, doing business with a social democratic government represented a significant departure from what they were accustomed to. In some respects, the traditional mechanisms of gaining access to the government were more restricted, or at least, perceived to be so. The sector strategy process represented a unique opportunity to develop contacts with the government and try to represent industry's point of view, or as some participants expressed it, to sensitize the NDP government to the realities of conducting business in the 1990s. Relatively few of the respondents were familiar with the Industrial Policy Framework, nor did they have significant views about it. For most of them, their involvement with the strategy development process was an outcome of their contacts with individual bureaucrats in ministries with sectoral leads or their involvement with sectoral associations. In some cases, participation in the sector strategies was largely for defensive reasons, to prevent the government from adopting policies they felt might harm the business community.

A limited number of participants had dealt with the government previously on specific programs, such as training initiatives, or targeted sectoral initiatives. For these participants, the strategy process was a valuable opportunity to continue to build on positive developments already underway. The decision to participate on the part of the unions was more straightforward. Several of the key industrial unions were already involved with sectoral training initiatives and saw the approach of the Industrial Policy Framework as a logical extension of that experience.[5]

Most of the strategies developed followed the format set out by government officials in the initial planning process. Although the direction of the strategies was left in the hands of the private sector participants, much of the analysis and writing was conducted by consultants to the government or by ministry officials. For the most part, this was viewed positively, although on occasion there was resentment of the excessive reliance placed on consultants to manage the process. In two of the sectors studied, Telecommunications and Plastics, the advisory committees were chaired by a prominent industry-based consultant and an academic respectively, both of whom were credited by the participants with making a substantial contribution to the successful outcome of the process.

The strategy development process varied widely from sector to sector and the resulting outcomes were equally diverse. This reflected the industrial structure of the different sectors, the strength and cohesiveness of sectorally based industry associations, their past history of sectoral activity, and the existence of sector-specific conflicts among individual firms within the sector, or, in one instance, regional clusters of firms. Differences were pronounced between highly regulated and more monopolistic sectors such as telecoms, which traditionally had considerable involvement with the federal government, and other sectors composed of smaller, more varied firms, such as the computing sector. In the telecom case, lingering conflicts over the recent federal decision to allow increased competition in the sector exerted a strong influence over the entire process. Some cases, such as plastics, had a long history of working together through their industry association, especially at the federal level and in other provinces. They had actually begun their efforts in Ontario before the SPF was formalized and quickly folded their ongoing activities into the sector development process. Other sectors were not really sectors per se. The health industries sector, for example, was assembled from four different subsectors – medical devices, pharmaceuticals, biotechnology and health services – with no prior history of working together. The cultural industries strategy also involved a diverse array of subsectors

– film, television, live drama, sound recording, and publishing assembled under one umbrella. Some respondents maintained that it felt like too broad a range of interests arrayed around the table to formulate a cohesive sectoral perspective.

Most of the sector strategies that were completed, identified a similar set of challenges confronting the sector – increased international competition and access to foreign markets were most frequently identified as the central issues. For many, the conclusion of the Canada–US Free Trade Agreement and the onset of the recession in the early 1990s accelerated the process of continental rationalization previously underway. This was particularly marked in sectors dominated by foreign ownership, such as the auto parts sector and the electrical and electronic manufacturing sector (Canadian Independent Automotive Components Committee, 1994; Electrical and Electronic Sector Advisory Council, 1995). Many of the smaller, lower value-added plants in these sectors produce similar products to those of sister plants in the US. The post-CUFTA and NAFTA trade environment placed a growing proportion of the Canadian plants in jeopardy, as the US parents tried to reduce costs by closing the less efficient plants. The challenge for the Canadian operations was to reduce their costs of production and increase the value-added of their output by raising skill levels and improving production processes and the organization of work within their operations. Even for those sectors in a stronger competitive position, such as the plastics industries, the issue of gaining an export market share in the US and reducing the trade deficit was viewed as a major challenge (Ministerial Advisory Committee on Plastics for the Province of Ontario 1994).

In other instances, such as the computing sector, the problems identified were quite different. The computing sector in Ontario is dynamic and was one of the few that continued to grow through the period of recession and industrial restructuring in the early 1990s. The sector is dominated over-whelmingly by relatively small firms run by entrepreneurial owner-managers. They face a number of challenges characteristic of this type of firm: barriers to access to capital (especially with regard to the question of valuing intellec-tual property assets as collateral), lack of adequate managerial skills, and a virtually unlimited need for more highly qualified employees, particularly graduates of the universities and colleges (Advisory Committee on the Computing Sector, 1993). The computing sector shared a concern with most of the other sectors about the barriers that limited its access to export mar-kets and its need to improve the export market development capabilities of its member firms through the creation of export consortia or better manage-ment training. The machinery and tool, die and mold sectors were also char-acterized by the presence of a large number of small and medium-sized enterprises. The firms in these sectors faced similar problems in terms of their capacity to adopt new technology, undertake training and develop new markets for their customers. The disincentive to invest in training due to the problem of poaching and the mobility of employees was also cited as a problem. The machinery sector, in particular, was characterized by a high degree of import penetration indicative of its structural weaknesses. The rising cost of investment in new technologies and the need for continuous skill upgrading were seen as major challenges for the sector (Machinery, Tool, Die, and Mold Sector Advisory Committee, 1996).

A closely related issue for many sectors involved the need for export market development. The key challenge for firms in these sectors was the expansion of trade opportunities abroad, especially in the larger continental market south of the border in the aftermath of the free trade agreement with the US. For instance, a key issue for firms in the electrical and electronics sector was the growing rationalization of foreign (especially US-) owned firms and the lack of global product mandates from their US parent firms (Electrical and Electronic Sector Advisory Council, 1995). Similarly, the plastics sector identified increased market penetration from the US and the trade deficit in resins as the key issue to be addressed. Given the relative size of the US market, moving to higher value-added products was seen as the most effective remedy for the trade deficit (Ministerial Advisory Committee on Plastics for the Province of Ontario, 1994). The health industries sector also identified the need to build export markets and acquire greater experience in the global marketplace as key challenges. These problems were compounded by the relatively small size of many of the firms in these sectors (especially the domestic firms), their absence of a strong connection to the research base in the province, the limited mandate for multinationals to conduct R&D in Ontario and the need for better market intelligence (Health Industries Advisory Committee, 1994).

Some sectors identified unique problems that they faced. For the auto parts sector, a key issue was their excessive reliance on selling to the Big Three US assemblers. Two significant developments in the auto industry were squeezing the parts manufacturers and putting added pressure on their competitive position. The first grew out of the trend on the part of the Big Three to shift more responsibility for product design and development onto their suppliers. The smaller size of Canadian firms in the sector relative to their American counterparts made it more difficult to develop and acquire needed production technology and to undertake the scale of product development demanded by the Big Three. One respondent pointed to the growing tendency of the auto assemblers to cut off any supplier who failed to meet the more demanding criteria and concentrate their supply base in a smaller number of firms. The second development was the growing presence of Japanese assemblers in North America and their tendency to buy from Japanese suppliers which also constrained the potential market for Canadian parts manufacturers (Canadian Independent Automotive Components Committee, 1994).

A key observation that emerges from the interviews is the radically different bases on which the various strategies were devised and implemented and the wide range of values and assumptions brought to the process by the respective participants. This is consistent with the point raised by Hollingsworth and Streeck – the pattern of relations among firms within a sector is influenced by the specific properties of that sector, and this pattern determines the social conditions under which firms must adapt to changes in technology and the broader economy. In a few instances, respondents referred to the exercise as an opportunity to get the sector to focus more on identifying its collective needs or to formulate more of a consensus vision about where the sector should be going. In a number of cases, sector participants identified a much broader awareness and understanding of the extent of technological changes underway and their potential impact on the sector as one of the main outcomes.

Despite the wide range of outcomes that emerged from the various strategies, most participants viewed the process itself as a success. In virtually all the cases studied, respondents felt that the sectoral analysis that was produced provided a sound analysis of the current strengths and weaknesses of the sector, as well as the opportunities and threats confronting it. This is not to say that the process was entirely without problems. In some sectors, traditional hostility between business and labor generated friction at the outset. In some instances, government facilitators were able to overcome the obstacles, while in one or two cases, key business or labor leaders broke the logjam. However, in at least one of the sectors studied, a number of the business participants were convinced that the strategy had been hijacked by the labor participants and some of them simply dropped out of the process after a certain point. The respondents questioned the representativeness of the resulting strategy. In another instance, internal conflicts between regional clusters of firms and different industry associations proved impossible to overcome, eventually prompting the chair of the advisory committee to resign in frustration.

In some sectors, some participants dismissed the entire process as merely a modified version of the venerable Canadian pastime of indulging in commissions and inquiries. In some sectors, more accustomed to dealing with government as a regulator of their activities, key participants viewed the process as just another government initiative – and an unnecessary one at that – rather than as an exercise in associative governance. However, even in the telecoms sector, where this view was most prevalent, several participants suggested that the strategy process had an impact. According to them, the creation of a 'vision' for the sector, on which the fractious participants could agree, was a major accomplishment. One participant maintained that another notable accomplishment was convincing the representative of a key firm in the sector to view itself not just as a telecom company, but as an infrastructure provider for the new economy.

In several sectors, the process clearly represented a novel and unique departure. In these cases, one of the most valuable outcomes was the strategy development process itself, in terms of establishing new relationships among potential suppliers and users and the effective identification of common sectoral interests. Some respondents indicated that their participation in the process helped increase their network of contacts within the sector and their awareness of other key players. For other respondents interviewed, one of the most beneficial outcomes was the increased visibility it generated for their sector – both internally in terms of the member firms, and in the eyes of the government bureaucrats. In the health sector, the strategy process represented an original attempt to look at the sector's role as an agent of economic development, rather than merely as a provider of services to government; one participant viewed the increased networking, especially between private sector suppliers and public sector purchasers, as a key outcome. In other sectors, such as computing, respondents indicated that the strategy process afforded them the opportunity to address a range of issues that they rarely discussed in their associations. One key participant in this strategy indicated that his involvement in the process provided him with a different view of how government worked and altered his view of how a sector could interact with government. The commitment of government support for approved sector strategies afforded the sectors the opportunity to launch initiatives that had been under discussion for years. Several respondents regarded the Connect–IT initiative that grew out of the computing sector strategy as a highly valuable development for the sector.

For the most part, the number of sectors where the strategy development process changed the organizational or business culture of the sector in any significant way appears to have been limited. Few of the respondents interviewed believed that the strategy process on its own fundamentally changed the organizational culture of their sector or contributed to a significant increase in the level of cooperation and trust among firms within the sector. In many sectors, the councils and other administrative bodies created did not outlast the defeat of the NDP government and the cancellation of the SPF in 1995. However, there were also significant exceptions to this rule – in the case of aerospace, the formation of an Ontario Aerospace Council as a direct result of the strategy development process was widely viewed as an innovative and much needed departure for the sector. While views varied within the cultural sector, for some, the creation of a Cultural Industry Council of Ontario was an important accomplishment. In the case of plastics, already one of the most cohesive and best organized sectors, many participants viewed the process as a success that fed directly into the ongoing process of consolidating the diverse elements of the sector into a more unified sectoral association. A number of respondents in specific sectors, health and plastics in particular, expressed their frustration with the failure of the Conservative government to follow through with the implementation of the strategies. One felt that the new government was proving to be very shortsighted in this regard. For the majority, however, the election of a

Conservative government in 1995 signaled a return to the traditional way of interacting with the provincial government. Indeed several respondents indicated that in the aftermath of the election, they reverted to lobbying a very sympathetic government around the more conventional macro issues of tax changes, labor relations and environmental regulation.

The high number of sectors that participated in the strategy development process would suggest that demands on the Sector Partnership Fund should have been high. Indeed there was no lack of recommendations for concrete initiatives in virtually all of the plans. These initiatives tended to be grouped into four categories: access to capital, technology and R&D, education and training, and export trade development. Despite this fact, the Sector Partnership Fund underspent its allocation in every year that it existed and at the time of its termination in July 1995, little more than half of the $150 million allocation had been committed. Of the concrete initiatives that received approval, the two largest went to sectors that had actually developed their strategies under the previous Liberal government. A number of factors accounted for this outcome. One that created considerable barriers was the expectation of substantial industry funding for the initiatives. The imposition of a 'quasi-market test' on SPF initiatives imposed a hurdle that many private sector participants had difficulty surmounting. One reason for this was that the firms who stood to benefit the most from the sector initiatives were the smaller and medium-sized ones with the least resources to fund them, whereas the larger firms in the sector often saw less of a need to contribute direct funding to the sector-specific initiatives. A related problem encountered was the bureaucratic approvals process required to access funding from the Sector Partnership Fund. In their concern to avoid creating another government pork barrel, central agency officials created a process that proved both intimidating and frustrating for sector participants and their lead ministry sponsors. The internal process broke down over an inability to surmount the traditional conflict between 'spenders' and 'guardians' inside the government. Many sectoral participants concluded in the end that whereas the strategy process successfully reflected the principles of associative governance, the approvals process failed miserably in this regard.

Despite these difficulties, a number of important initiatives emerged from the process: the establishment of permanent advisory committees such as the Ontario Aerospace Council and the Cultural Industry Council of Ontario and the Council for an Ontario Information Infrastructure (including members from both the telecom and computing sectors); providing the government with better access to sector leaders for consultation; developing a more efficient way of managing relations between sectors and the government as one sector leader noted above; and using the SPF as a lever to bring sectoral partners together to assess the competitive position of their sectors and formulate cooperative strategies to respond to the challenges they faced. In this respect, the sector strategy process bore a striking resemblance to the program of drafting Regional Technology Plans (later Regional Innovation Strategies) launched by the European Union in 1994 (Landabaso, Oughton and Morgan, 1999; Morgan and Nauwelaers, 1999).

In general, the initiatives put forward for approval were grouped into four categories: access to capital, technology transfer and R&D, education and training, and export market development. Several significant initiatives were approved and received funding from the Sector Partnership Fund. A number fell into the second category oriented towards the creation of sector-based technology centers. Examples included: the Guelph Food Technology Center, designed to increase effective technology and information transfer, as well as to provide accessible pilot plant facilities for the food industry; an Ontario Center for Environmental Technology Advancement to provide technical support services, financial advice and business counseling to help young firms commercialize environmental technologies; Connect-IT, a computing Sector Resource Facility to assist the many small and medium-sized firms in Ontario's industry in developing sector-specific competency in management, standards, marketing expertise and export readiness. In the computing sector, funding was also provided to support the Electronic Commerce Institute to promote the adoption and use of electronic data interchange in Canadian industry. Other areas that received some funding included export market development through Interhealth Canada, a private, not-for-profit corporation designed to pursue and gain international contracts for Canadian firms in key markets around the world; and the plan to establish representatives for the auto parts sector in Japan and Europe to help increase sales to Japanese and European assemblers in their North American and foreign operations. Several sectors suffered from the fact that their strategies were completed and recommendations formulated just as the new Conservative government terminated funding for the Sector Partnership Fund in July, 1995.

Although the SPF was the main funding source for the sector strategies, a number of recommendations were acted upon through other government initiatives. One of the largest single programs to emerge from the strategy development process – the $100 million Ontario Network Infrastructure Program – was actually financed out of the government's capital budget and fell outside the parameters of the SPF. This program played a vital role in funding community-based networks and providing increased access to the internet across the province and was followed with the broad vision of the telecom sector strategy. Another outcome that emerged from a key recommendation of both the computing and telecom strategies was a revised government process to allow for Common Purpose Procurement in 1995. Many of the strategies made recommendations to improve the quality of training in their sector, but the primary agency responsible for training in the province was also abolished after the election (Wolfe, 1997).

Conclusion

The task of evaluating the sector strategy process in Ontario is made more difficult by the extreme change in policy direction that followed the election of the Conservatives in 1995. Virtually all of the momentum and most of the specific initiatives developed under the NDP government were abandoned. Although the strategy development process succeeded in drawing a larger number of participants to the table across a broader range of sectors, in terms of concrete outcomes, it differed little from other similar exercises conducted by the federal and provincial governments in the past, as several of our respondents suggested.

There are relatively few sectors that began to move towards the establishment of a 'negotiated order' in the sense implied by Moore and Booth. Those sectors where this occurred to some extent seem to be ones where the process built upon a solid base at the outset – such as plastics, which already enjoyed a strong degree of effective internal organization and had already signed a Memorandum of Understanding with the government prior to the creation of the Sector Partnership Fund. In the case of the computing sector the strategy process helped build more of a collective identity in what had previously been an excessively fragmented sector. This proved to have some lasting benefit, both through the establishment of an Ontario wing of the national association and the creation of Connect-IT.

The limited nature of the changes effected by the sector strategies may very well be a result of the broader configuration of social and political forces in provincial politics. The shift in policy regime envisaged by the Industrial Policy Framework and the launching of the sector strategies were based upon the new electoral realities after 1990. For the sector approach to have effected a more significant change in the business culture of the province, those realities needed to be confirmed to some degree in the subsequent election. The reversal of political climate in June, 1995 precluded that possibility. Had either of the other two parties won the election, the Sector Partnership Fund would most likely have continued and the results of the strategy development process been extended. As Neil Bradford has perceptively argued, the success of experiments in associative governance depend upon the mix of party politics, state capacity and societal interests. In the Ontario case of the early 1990s, partisan factors were responsible for both the initiation and the termination of the experiment; the internal bureaucratic structures of the state responded more successfully to the challenge of devolving responsibility for the strategy development process than they did with the approvals process; and ultimately, the experiment left little lasting impact on the organization of business interests and the business culture in the province. 'In liberal polities like Ontario . . . the prospects for robust associative innovation depend on the incentives for business to cooperate, or at least, not to exercise its option to "exit" the partnership' (Bradford, 1998,

p. 541). The partisan outcome of the 1995 election effectively removed those incentives.

In some nascent sectors, such as digital media and biotechnology, the Conservative government has preserved or re-established a modified version of a sectoral approach. Thus, it may still be too early to judge the final legacy of Ontario's experiment in associative governance. For some sectors, the experience of working together was a novel and satisfying one; the perceived benefits may continue to trickle down to their members for some time to come. For others, the strategy development process reinforced a trend towards internal organization that existed previously and the process of developing internal networking mechanisms may continue. Although few sectors approached the level of a negotiated order, the future of the approach bears close watching.

Notes

1 Financial support for the research in this paper was provided by SSHRCC Research Grant No. 809–95–0009. The paper draws, in part, on research conducted jointly with Meric Gertler. I would like to thank Shauna Brail, Sean DiGiovanna, Bob Marshall, Ammon Salter and Mikael Swayze for their help in the conduct of the interviews and Shauna Brail and David Garkut for their assistance in analyzing the results. I am deeply indebted to the more than one hundred participants in the sector strategies who took time from their busy schedules to respond to our questions. I am grateful to Bill Coleman and Michael Keating for their helpful comments on the initial draft of this paper. The account that follows is necessarily biased, given my involvement in some of the activities described as the Executive Coordinator for Economic Development in the Cabinet Office, Government of Ontario from 1990 to 1993. Responsibility for any remaining errors or omissions is mine alone.
2 A more detailed account of the political challenges faced by the NDP government of Ontario can be found in Rachlis and Wolfe 1997. An overview of the key features of Ontario's innovation system are also presented in Wolfe and Gertler, 1998.
3 For a fuller discussion of the evolution of industrial and technology policy in Ontario during the 1980s and 1990s, cf. Wolfe, 1999.
4 Each interview covered the same basic issues: the reasons for involvement with the strategy and the issues confronting the sector; the mechanics of the strategy development process and their effectiveness; and the outcomes of the process in terms of both specific recommendations and the initiatives that flowed from them, as well as the impact of the process on broader relations within the sector.
5 For a detailed discussion of the government's training agenda, cf. Wolfe, 1997.

References

Advisory Committee on the Computing Sector (1993) *Agenda for Action: a Strategy for the Development of Ontario's Computing Sector*, Toronto.

Amin, A. (1996) 'Beyond Associative Democracy', *New Political Economy*, Vol. 1, No. 3.

Atkinson, M.M. and W.D. Coleman (1989) *The State, Business, and Industrial Change in Canada*, The State and Economic Life Series, Toronto: University of Toronto Press.

Bradford, N. (1998) 'Prospects for Associative Governance: Lessons from Ontario, Canada', *Politics and Society*, vol. 26 (4), (December), pp. 539–73.

Canadian Independent Automotive Components Committee (1994) *Strategic Action Plan*, Toronto.

Coleman, W.D. (1988) *Business and Politics: a Study in Collective Action*, Montreal and Kingston: McGill-Queen's University Press.

Electrical and Electronic Sector Advisory Council (1995) *The Ontario Electrical and Electronic Sector Strategy*, Toronto.

Gourevitch, P. (1986) *Politics in Hard Times: Comparative Responses to the International Economic Crisis*, Ithaca and London: Cornell University Press.

Gregersen, B. and B. Johnson (1997) 'Learning Economies, Innovation Systems and European Integration', *Regional Studies*, vol. 31 (5), pp. 479–90.

Hall, P.A. (1993) 'Policy Paradigms, Social Learning and the State: the Case of Economic Policymaking in Britain', *Comparative Politics*, April, pp. 275–96.

Health Industries Advisory Committee (1994) *Healthy and Wealthy: a Growth Prescription for Ontario's Health Industries*, Toronto.

Hollingsworth, J.R. and W. Streeck (1994) 'Countries and Sectors: Concluding Remarks on Performance, Convergence and Competitiveness', in J.R. Hollingsworth, P.C. Schmitter and W. Streeck (eds), *Governing Capitalist Economies: Performance and Control of Economic Sectors*, New York and Oxford: Oxford University Press.

Landabaso, M., C. Oughton and K. Morgan (1999) 'Learning Regions in Europe: Theory, Policy and Practice through the RIS Experience', Paper Presented to the 3rd International Conference on Technology and Innovation Policy, Austin, TX, 30 August–2 September.

Laughren, F. (1991) 'Ontario in the 1990s: Promoting Equitable Structural Change', Budget Paper E, in *1991 Ontario Budget*, Toronto: Queen's Printer for Ontario, 29 April.

Lundvall, B.-Á and S. Borrás (1998) *The Globalising Learning Economy: Implications for Innovation Policy*, Targeted Socio-Economic Research Studies, DG XII, Commission of the European Union, Luxembourg: Office for Official Publications of the European Communities.

Machinery, Tool, Die and Mold Sector Advisory Committee (1996) *Retooling for Competitiveness: Machinery, Tool, Die and Mold Sector Strategy*, Toronto.

Majone, G. and A. Wildavsky (1984) 'Implementation as Evolution (1979)', J.L. Pressman and A. Wildavsky, in *Implementation*, 3rd. ed., expanded, Berkeley: University of California Press.

Ministerial Advisory Committee on Plastics for the Province of Ontario (1994) *Ontario's Plastics Industry: a Winning Strategy*, Toronto.

Moore, M. and S. Booth (1989) *Managing Competition: Meso-Corporatism, Pluralism and the Negotiated Order in Scotland*, Oxford: Clarendon Press.

Morgan, K. (1999) 'A Regional Perspective on Innovation: from Theory to Strategy', in K. Morgan and C. Nauwelaers (eds), *Regional Innovation Strategies: the Challenge for Less Favoured Regions*, London: Stationery Office.

Morgan, K. and C. Nauwelaers (eds) (1999) *Regional Innovation Strategies: the Challenge for Less Favoured Regions*, London: Stationery Office.

Nelson, R.R. and S.G. Winter (1982) *An Evolutionary Theory of Economic Change*, Cambridge, MA: Belknap Press.

Ontario, Ministry of Economic Development and Trade (1995) *Ontario Sector Snapshots*, a progress report on the sector development approach, Toronto.

Ontario, Ministry of Industry, Trade and Technology (1992) *An Industrial Policy Framework for Ontario*, Toronto: Queen's Printer for Ontario.

Pontusson, J. (1995) 'From Comparative Public Policy to Political Economy: Putting Political Institutions in Their Place and Taking Interests Seriously', *Comparative Political Studies*, vol. 28 (1), April, pp. 117–47.

Rachlis, C. and D. Wolfe (1997) 'An Insider's View of the NDP Government of Ontario: the Politics of Permanent Opposition Meets the Economics of Permanent Recession', in G. White (ed.), *The Government and Politics of Ontario*, 5th. ed., Toronto: University of Toronto Press.

Sabel, C.F. (1992) 'Studied Trust: Building New Forms of Co-Operation in a Volatile Economy', in F. Pyke and W. Sengenberger (eds), *Industrial Districts and Local Economic Regeneration*, Geneva: International Institute for Labour Studies.

Streeck, W. and P.C. Schmitter (1985) 'Community, Market, State – and Associations? The Prospective Contribution of Interest Governance to Social Order', *European Sociological Review*, vol. 1 (2), September: pp. 119–38.

Walkom, T. (1994) *Rae Days: the Rise and Follies of the NDP*, Toronto: Key Porter Books.

Wolfe, D.A. (1997) 'Institutional Limits to Labour Market Reform in Ontario: the Short Life and Rapid Demise of the Ontario Training and Adjustment Board', A. Sharpe and R. Haddow (eds), in *Social Partnerships for Training: Canada's Experiment with Labour Force Development Boards*, Kingston: Centre for the Study of Living Standards, Caledon Institute of Social Policy and School of Policy Studies, Queen's University.

Wolfe, D.A. (1999) 'Harnessing the Region: Changing Perspectives on Innovation Policy in Ontario', in T.J. Barnes and M.S. Gertler (eds), *The New Industrial Geography: Regions, Regulation and Institutions*, London: Routledge.

Wolfe, D.A. and M.S. Gertler (1998) 'The Regional Innovation System in Ontario', H. Braczyk, P. Cooke and M. Heidenreich (eds), in *Regional Innovation Systems: the Role of Governances in a Globalized World*, London: UCL Press.

Index